Power and Influence

Power and Influence

The Metaphysics of Reductive Explanation

RICHARD CORRY

OXFORD
UNIVERSITY PRESS

OXFORD
UNIVERSITY PRESS

Great Clarendon Street, Oxford, OX2 6DP,
United Kingdom

Oxford University Press is a department of the University of Oxford.
It furthers the University's objective of excellence in research, scholarship,
and education by publishing worldwide. Oxford is a registered trade mark of
Oxford University Press in the UK and in certain other countries

© Richard Corry 2019

The moral rights of the author have been asserted

First Edition published in 2019
Impression: 1

Published in the United States of America by Oxford University Press
198 Madison Avenue, New York, NY 10016, United States of America

British Library Cataloguing in Publication Data
Data available

Library of Congress Control Number: 2019943599

ISBN 978-0-19-884071-8

Printed and bound in Great Britain by
Clays Ltd, Elcograf S.p.A.

Contents

Acknowledgements

This book has taken much longer to complete than expected, and it attempts to bring together ideas that I had been developing for many years before I began work on this manuscript. Thus, many people have contributed one way or another. However, there are a number of people who deserve special thanks. First and foremost, I am indebted to my family, Pragati, Nishka, and Mira, for their unfailing support and encouragement. Thanks to David Coady for providing a critical, but encouraging sounding board for my ideas over the past fourteen years; to Huw Price and Toby Handfield for giving helpful feedback on the initial germs of the view developed here; to Roger Latham for discussions of the nature of science during our runs through the wilderness of Tasmania; and to Janny Corry for being willing to read and edit the first rough draft. I would also like to thank two anonymous referees who provided extensive feedback on an earlier draft of this book. Although I doubt that I have adequately addressed all of their concerns, I am sure that this is a much better book due to their input. The School of Humanities here at the University of Tasmania provided a period of study leave that was essential for finally bringing my ideas together into a coherent whole.

Parts of Chapter 3 are reprinted from my paper "How is Scientific Analysis Possible?" in *Dispositions and Causes* (2009) edited by Toby Handfield, Oxford University Press, Oxford. I am grateful to Oxford University Press for kind permission to reuse this material here.

List of Figures

1
Introduction

Here is a fun experiment. Take a bucket of water to a first-floor balcony or some similar high place (you will need a drop of at least 2m to the ground). Now fling the water out of the bucket whilst swinging the bucket sideways. With just a little practice you will find that you can produce a smooth sheet of water that falls towards the ground. But here is the cool bit: after falling for just under a second the sheet of water will spontaneously and mesmerizingly disintegrate into hundreds of small droplets.

Why does the water behave like this? Here is an explanation: water is made up of molecules that have a weak attraction for each other. This attraction is what holds the molecules together to form the initial sheet of falling water. However, the air through which the water is falling is also made up of molecules, and as the water falls through the air, there are collisions between the molecules of air and the molecules of water. Whilst the water is moving slowly, these collisions are not energetic enough to overcome the attraction between the water molecules, but the water is accelerating as it falls and when it reaches a critical velocity, the collisions between air and water molecules become energetic enough to tear the water molecules away from each other. At this point the sheet of water spontaneously disintegrates.

This explanation of the water's behaviour is, admittedly, rather sketchy, but it nonetheless provides some understanding of the phenomena in question. If we had more detailed information about the nature of the parts involved and their interactions, then we could provide more detailed explanations. So, for example, if we knew the masses of the air and water molecules, the density of their distribution in the relevant area, the strength of the attractive forces between water molecules, and the strength of Earth's gravitational attraction, then we could explain why it takes just under one second for the sheet of water to disintegrate. What is more, with this kind of information we could make predictions about what would happen if things were slightly different. What would happen if the experiment were performed in an area of lower gravitation? What if it were performed in a vacuum?

The explanation above is an example of what I will call *reductive explanation*[1]. Roughly speaking, a reductive explanation proceeds by treating the system of

---

[1] I use this term somewhat hesitantly, as there are many different understandings of 'reduction' in the literature. The particular way I understand the term will be made clear in Chapter 2.

interest as being made of component parts, and the phenomenon is then explained by reference to the properties of, and interactions between, those parts. In our example, the water and air are treated as being made up of particles, and the behaviour of the system is explained using our knowledge of how these particles behave on their own, and in pairs (pairs of water molecules attract each other; a water molecule and a particle of air exchange momentum when they collide). In this case, then, the relevant components are not just the individual water molecules and air particles, but also systems consisting of pairs of such entities. Thus, the explanation proceeds by treating the system as made up of smaller objects (water molecules and particles of air), and then using our knowledge of the behaviour of one- and two-object subsystems in order to explain the behaviour of the whole system.

Recently, reductive explanation has acquired a bad reputation. In the life sciences, for example, there has been a focus on the limits of reductionism with some calling for a move 'away from reductionism and toward a new kind of biology for the 21st century' (Elser and Hamilton, 2007, 1403). In everyday discourse, 'reductionist' is often used almost interchangeably with 'simplistic' and has become a term of abuse (e.g. 'Victim-blaming the left is so reductionist and dumb' (Wilson, 2016)). Reductionism is sometimes even equated with an oppressive political ideology as in this blog post:

> 'So reductionist "science" did not even have to fight too much for its absolute domineering, suppressive, dis-empowering paradigm—it fitted the political and military powers-to-be perfectly and was supported by major funding from both. The net outcome: anti-life "Franken-science" killing the whole planet, not only humans, by geoengineering, GMO, vaccines etc'. (Maksimovich-Binno, 2015)

As Richard Dawkins puts it, reductionism is 'one of those things, like sin, that is only mentioned by people who are against it' (1986, 13).

This negative view of reductionism is unfortunate, for the reductive method is one of the most powerful tools available to science. By studying fairly simple systems in relative isolation, and then applying our knowledge of these systems to explain and predict the behaviour of more complex systems, we have a way to gain a foothold in our understanding of systems that may be too complex to get a handle on if studied directly. The method also allows us to extend our understanding to new systems, and to unify our understanding of different systems. The attraction between water molecules, for example, can also be used to explain the almost spherical shape of the droplets that form when the sheet of water disintegrates and even how it is that some insects are able to walk on the surface of water. Even if you do not agree with Friedman (1974) and Kitcher (1989) that unification is the ultimate basis of scientific explanation, you must surely agree that such explanatory unification should be regarded as a Good Thing. Furthermore, because the analytic method allows us to predict the behaviour of such a large range of systems based on a unified theory of the components of those systems, this unified theory is capable of much more rigorous experimental testing than would be possible for theories that did not analyse the complex systems, and so did not refer to commonalities between them. Finally, it is this same reductive method that allows us to put together complex systems from parts that we understand in order to produce useful and predictable technologies.

As the physicist Paul Davies points out:

> Few would deny the efficacy of the reductionist method of investigation. The behaviour of gases, for example, would lack a satisfactory explanation without taking into account the underlying molecular basis. If no reference were made to atoms, chemistry would amount to little more than a complicated set of ad hoc rules, while radioactivity would remain a complete mystery. (2006, 35)

Davies tells us that 'the jewel in the crown of reductionist science is subatomic particle physics, with its recent extension into superstring theory and M theory' (2006, 35), but the reductionist method is not confined to physics; it is ubiquitous in the sciences and in everyday life. Chemists explain the action of acids in terms of the interactions of individual H^+ and OH^- ions. Evolutionary biologists explain the distribution of traits in a population in terms of the interactions of the individuals bearing those traits. Political analysts explain and attempt to predict election outcomes in terms of the voting preferences of groups of voters. Economists explain changes in the market by referring to the buying and selling preferences of individual agents.

Recent critiques of reductionism, however, do make one thing clear: the success of the reductive method is not guaranteed. In any given situation, the world could have been such that the method would not work. Thus, the success of the reductive method in such diverse fields suggests that the method really has latched onto a substantive fact about the way that the world is structured.

The aim of this book, therefore, is to consider the question 'What must the world be like in order for the reductionist methodology to succeed?' I begin, in Chapter 2, by examining the nature of reductive explanation. In Chapter 3, I argue that the reductionist approach involves a number of assumptions about the nature of properties and the way they interact. Most importantly, when we use our knowledge of how objects behave individually or in pairs to explain the behaviour of a system consisting of many such objects, we are assuming that what we know about the subsystems still applies (in some sense) when those subsystems form part of a more complex system. This assumption, in turn, implies that something remains constant, or invariant, in the move from the well-understood simple situation to the complex situation that we hope to explain. I show that the kind of invariant required for reductionist explanation cannot be found among the obvious candidates in existing ontologies, and in Chapters 3 and 4, I suggest that the invariance required by reductionist explanation is best supplied by an ontology of causal powers that exert influences which are directed towards particular outcomes—an ontology of power and influence. Having described the ontology of power and influence, I return once more to the nature of reductive explanation in Chapter 5, where I develop a detailed model of the reductive method in terms of the new ontology of power and influence. One of the main aims of this chapter is to make explicit the metaphysical assumptions regarding this ontology that are built into the reductive approach.

The remainder of the book puts the ontology of power and influence to use to shed light on a number of questions in the metaphysics of science. In Chapter 6, I respond to Alexander Bird's argument that we have no good reason to believe in macroscopic causal powers. I show that the ontology of power and influence can be extended from the micro-realm to the macro, and thereby support my claim that the reductive method is carried out at all levels of explanation. In Chapter 7, I show that an ontology of power and influence can serve as a ground for the laws of nature (where I take 'laws' to be general statements of a particular kind, rather than the truthmakers of such statements), and show that this ontology solves some of the problems involved with attempts to ground the laws of nature in a more traditional ontology of causal powers. In Chapter 8, I consider the relation between power, influence, and causation, and in Chapter 9, I suggest a modification to the techniques of causal modelling which will, on the one hand, make the underlying metaphysics clearer and, on the other hand, will make the technique more powerful by showing how existing causal models can be extended to cover new situations. In Chapter 10, I consider what happens when the assumptions of the reductive approach are not satisfied. I argue that the failure of some of these assumptions would give rise to what could legitimately be called 'emergent' phenomena. In this way, I suggest, the ontology of power and influence can be used to give a clear, positive account of emergence, and to show that two distinct kinds of emergence are possible. Finally, in Chapter 11, I argue that the ontology of power and influence may shed light on issues beyond metaphysics. In particular, I argue that this ontology suggests a new class of theories in normative ethics, which I call *influentialism*. I do not claim that influentialism is better than all its rivals, but I do argue that it has its virtues and is worth considering.

Before moving on to the substance of the book, there is one more point I should make. In some instances it may be useful to distinguish the concept of a *reductive explanation* from the *reductive method* for developing, justifying, and extending such explanations. In the reductive method, one studies component systems in relative isolation, and then uses what one learns in this way to explain the behaviour of a complex system made up from these components. But note that it is possible to develop a reductive explanation without making use of the reductive method. This happens when one explains the behaviour of a system by *hypothesizing* that the system is made up of components that behave in a certain way. The existence of atoms with various properties, for example, was hypothesized for the purpose of reductive explanation long before atoms could be studied in isolation. The justification for a reductive explanation in such a case comes from its predictive and unificatory power.

Whether a reductive explanation has been developed by the reductive method or by hypothesis, we often use a variation of the reductive method to extend our explanations to cover new systems, or new phenomena. To do this, we take what we know about the behaviour of the components in one system and use this to explain the behaviour of some other system containing similar components. In general, then, the reductive method proceeds by gaining an understanding of component systems in one circumstance and employing this understanding to explain the behaviour of a complex system which contains such components, though in circumstances that may be quite different from that in which they were originally understood. In most of this book I will not distinguish between the 'method' of attempting to provide a reductive explanation and the reductive method of developing, justifying, or extending such an explanation. However, this distinction will be important in Chapters 5 and 10.

2

Taking Apart the World

In this book I will be arguing that the reductive method, or at least one version of it, involves a number of metaphysical assumptions about the structure of the world. Before diving into the metaphysics of reduction, however, it is necessary to get a bit more specific about the reductive method that I have in mind, for there are a number of different things that one might mean by 'reduction'. In this chapter, then, I will outline the relevant notion of reductive explanation, and contrast it with other, related types of explanation.

2.1 Explanatory Reduction vs Theory Reduction

The contemporary philosophical literature on reduction in science was initiated by Ernest Nagel (1949; 1961), and is dominated by a concern with what is now called *theory reduction*. As the name suggests, theory reduction is concerned with a relation between theories. Nagel considers a theory to be a collection of statements, at least some of which are laws of nature. Theory T_B is then said to reduce to theory T_A if and only if T_B is derivable from T_A. So, for example, a biological theory would reduce to atomic physics if all the laws and claims of the biological theory could be deduced from the laws and claims of atomic physics. One problem for this account of theory reduction is the fact that T_A and T_B may have very different vocabularies, and may divide the world up in quite different ways. Atomic physics, for example, does not mention organisms, or genes, or fitness. In general, if T_A has a different vocabulary from that of T_B, then the laws of T_A cannot logically entail the laws of T_B on their own. Nagel gets around this problem by modifying his account of reduction to state that theory T_B reduces to theory T_A if and only if T_B is derivable from T_A together with a set of so-called *bridge laws*, or *coordinating definitions*, that translate the vocabulary of the high-level theory into the vocabulary of the low-level theory. As Nagel puts it, 'A reduction is effected when the experimental laws of the secondary science (and if it has an adequate theory, its theory as well) are shown to be the logical consequences of the theoretical assumptions (inclusive of the coordinating definitions) of the primary science' (1961, 352). One of the most successful examples of theory reduction is the reduction of thermodynamics to statistical mechanics. Here we have clear bridge laws such as that which states that the term 'heat' in thermodynamics can be translated as 'mean molecular kinetic energy' in statistical mechanics.

Needless to say, Nagel's account of theory reduction has faced much criticism. The most famous is the 'Multiple Realization Argument' advanced by Hillary Putnam (1967), Jerry Fodor (1974), and others. This argument attacks the coordinating definitions that are required by Nagelian reduction. Putnam and Fodor point out that many of the properties that are discussed in the high-level sciences are realized by a variety of distinct properties at the lower level. For example, Gresham's Law in economics tells us what happens in monetary exchanges under certain conditions. But a monetary exchange can clearly be realized in many ways–by handing over dollar bills, by signing cheques, or by changing the numbers in a computer memory. If a high-level property is multiply realized in this way, then it cannot be identical with any particular low-level property, and so cannot be translated into the language of the low-level science in the way required for Nagelian reduction. What is more, in such cases of multiple realization, the various low-level descriptions may miss the regularities that are found using the high-level vocabulary, and so may miss the explanations available at the high level.

A second complaint is that Nagel's account does not allow for the possibility of correction to the reduced theory in light of the reduction (see, for example, Feyerabend, 1962; Churchland, 1986; Schaffner, 1993; Bickle, 1998). For example, there is a clear sense in which the success of Newton's theory of gravitation can be explained by Einstein's theory of general relativity, and so we might be inclined to say that Newton's theory can be reduced to Einstein's. But Newtonian gravitation cannot be deduced from general relativity: all that can be deduced is that Newton's theory will be approximately correct in most everyday circumstances. Thus, Newtonian gravitation does not reduce, in the Nagelian sense, to general relativity. For this reason, we may be inclined to look for an alternative account of reduction which allows for less than complete deducibility.

More recently, philosophers and scientists have objected that the Nagelian account's focus on the reduction of entire theories seems to miss most of the 'reductive' relations that actually interest scientists. As the biologists Fang and Casadevall note, 'Exploring the epistemic relationships between different disciplines might be grist in the mill for a philosopher of science but does not seem a particularly fruitful endeavour for a working scientist' (2011, 1401). Many reductive relations that one finds discussed in science are *local* and *partial* reductions (see Schaffner, 2006, 397–8; Wimsatt, 2006, 448; Wimsatt and Sarkar, 2006, 697; Gillett, 2016). Rather than reducing an entire theory to another, these reductions focus on explaining how a small range of phenomena at one level of description can be explained in terms of a lower-level theory.

In this book, I am not concerned with theory reduction, but with these local, partial reductions that are the stock in trade of working scientists. Following Brigandt and Love (2017), let us refer to these local reductions as *explanatory reductions*. Explanatory reduction differs from theory reduction in two important ways. First, the relata of an explanatory reduction need not necessarily be theories: they may be fragments of a theory, generalizations of varying scope, or even particular facts. So, for example, the explanation of the behaviour of falling water in the Introduction to this book does not provide a reduction of one scientific theory to another. Rather, it shows how a restricted generalization (water falling near the Earth tends to form a sheet which spontaneously disintegrates into droplets) follows from certain facts that are typical of such situations (e.g. facts about the atmosphere and gravitational field near the Earth's surface) together with some more general claims that might count as theory fragments (e.g. water molecules attract).

The second characteristic difference between explanatory reduction and theory reduction is that Nagalian theory reduction is 'formal'—focusing on logical or semantic-analytical relations between (sets of) statements, while explanatory reduction is 'substantive' (Sarkar, 1998) or 'ontic' (Kaiser, 2015) in the sense that it focuses on properties and relations that exist *in the world*. Furthermore, like many authors on the topic of reductive explanation (see, for example, Kauffman, 1970; Wimsatt, 1976; Hüttemann and Love, 2011), I am concerned with *causal* explanations; I am not concerned, for example, with reductions that explain the shape of a macroscopic object in terms of the shapes and distance relations between its microscopic parts. Consider the explanation I gave earlier for the behaviour of falling water. This explanation refers to the properties of, and causal interactions between, molecules of water and air, and so is concerned with the properties and relations of entities in the world rather than relations of logical entailment between theories.

From here on, when I refer to reductive explanation, I will have in mind causal-explanatory reduction rather than theory reduction.

2.2 Characterizing Explanatory Reduction

What, then, distinguishes reductive explanation from other types of (substantive, ontic) explanation? It is a common thought that the characteristic feature of reductive explanations is that they explain properties or processes of a complex whole by reference to the properties of, and interactions between, its parts (Love and Hüttemann, 2011; Bechtel and Richardson, 2010; Sarkar, 1998; Wimsatt, 1976). In the Introduction to this book, for example, I explained the behaviour of a body of falling water by reference to the interactions between the water molecules from which it is composed. Note, however, that this explanation also made reference to the interactions between water molecules and molecules of air, which are clearly not parts of the body of water whose behaviour is being explained. Many putatively reductive explanations make reference to background conditions, or interactions with parts of the environment. Consider the following, seemingly reductive explanation of the process of muscle contraction: 'When signaled by a motor neuron, a skeletal muscle fiber contracts as the thin filaments are pulled and then slide past the thick filaments within the fibe's sarcomeres' (OpenStax, 2016, §10.3). The thin and thick filaments mentioned here are parts of the muscle fibre, but the motor neuron is external to the fibre. Similarly, Levins and Lewontin tell us that in ecology 'reductionism takes the form of regarding each species as a separate element existing in an environment that consists of the physical world and of other species' (Levins and Lewontin, 1980, 49).

If we are to classify these explanations as reductive, then reductive explanations cannot focus exclusively on the parts that make up an entity whose behaviour is being explained. For this reason, a number of authors have characterized reductive explanations in terms of a more abstract hierarchy of 'levels of organization' instead of part–whole relations. In a recent book on the topic of reductive explanation in biology, for example, Marie Kaiser argues for three characteristics of reductive explanation. According to Kaiser, a reductive explanation in biology is one which explains the behaviour of a biological object or system Y by:

(1) referring only to factors that are located on a lower level of organization than Y,

(2) focusing on factors that are internal to Y, that is, that are biological parts of Y, and

(3) representing the biological parts of Y as if they were parts in isolation (i.e., representing only those relational properties of and interactions between the parts that can be studied in other contexts than in situ).

(Kaiser, 2015, 235)

Characteristics (1) and (2) here serve as generalizations of the claim that reductive explanations are part–whole explanations (I will consider characteristic (3) shortly). Characteristic (2) tells us that reductive explanations focus on parts internal to the system being explained, but she says that this is merely a typical feature of reductive explanation, not a necessary one, since a reductive explanation for the behaviour of system Y may refer to factors in the environment of Y.

Characteristic (1), on the other hand, is taken by Kaiser to be a necessary feature of reductive explanation. If an explanation is to count as reductive, she believes, then the factors that are referred to in the environment must, like the parts of Y, be located on a lower level of organization than Y. Of course, to make full sense of this claim we need an account of levels of organization. Kaiser tells us that an object, X, is located on a lower level than object, Y, if and only if either X is a biological part of Y, or X belongs to the same general biological kind as one or more of the biological parts of Y.

Now, my ultimate aim in this chapter is to develop a general characterization of explanatory reduction that can be applied in any context. Kaiser's characterization, on the other hand, is clearly specific to biological explanation, since it makes central reference to the concepts of *biological kind* and *biological part*. Now it may be possible to generalize the concepts of biological kind and biological part so that they apply in explanations beyond biology. Carl Gillett (2016), for instance, has provided a careful general characterization of the circumstances under which scientific explanations tend to treat one entity as a part of some more complex entity, and this concept of a part might do the job we need. However, I do not think that it is necessary in our case to look for such a generalization, for I do not believe that we should restrict our concept of reductive explanation by insisting that such explanations cannot refer to higher-level environmental factors.

Consider, again, the explanation given above for the behaviour of falling water. This explanation refers to the parts from which the water is composed and the parts from which the air around the water is composed. But the explanation also refers to the gravitational attraction of the particles of water to the Earth. Clearly, the Earth is not an object at a lower level of organization than the body of falling water. Indeed, the Earth has parts that are themselves bodies of water, and so the obvious generalization of Kaiser's notion of levels of organization would put the Earth at a higher level of organization than the falling water. Nonetheless, the explanation for the behaviour of the falling water seems to be a reductive explanation. Certainly, it is a paradigmatic example of the kind of explanation that I am interested in here. Reductive explanations, as I understand them, can refer to environmental factors at any level of organization.

At least for my purposes in this book, then, I will avoid characterizing reductive explanations as part–whole explanations, and also avoid characterizing reductive explanations in terms of levels of organization. Instead, I will replace Kaiser's characteristics (1) and (2) above with the claim that a reductive explanation is one in which some behaviour within a complex system Y is explained by reference to the properties and interactions between parts of Y. The behaviour to be explained could be a behaviour of the system as a whole (in which case we have a straightforward part–whole explanation), or it could be the behaviour of some proper subset of the parts of Y. The explanation of the the behaviour of falling water, for example, has the latter form, since the complex system of interest contains water molecules, air molecules, and the Earth as basic parts, but the body of water whose behaviour is being explained is a collection of some, but not all of these parts. The explanation of muscle contraction also has this form, since the basic parts include the motor neuron, the thin fibres, and the thick fibres, while the muscle is composed of the thin and thick fibres but not the motor neuron.

So Kaiser's notion of a biological part, and Gillett's more general discussion of parthood, are not relevant here. I am concerned with explanations that refer to interactions between the parts of complex systems, and I intend that my discussion should apply both in cases like those Kaiser and Gillett discuss, where the complex system is regarded as an individual (such as an ion channel or a diamond), and in cases where the system is not regarded as a unified entity.

So far, my characterization of reductive explanation is broader than Kaiser's, which is good, since I want it to apply beyond biology. But one might be worried that it is too broad, that it does not capture the *reductive* nature of reductive explanation. It is here that Kasier's characteristic (3) comes in to play. It is this third characteristic which captures what many people object to in reductive explanations (see the discussion at the beginning of Chapter 3), and it is this third characteristic that is of most interest so far as this book is concerned.

As suggested by Kaiser's parenthetical gloss, characteristic (3) has as much to do with the *methodology* of reductive explanation as it has to do with the form of such explanations. When constructing a reductive explanation for a behaviour within some system, Y, we take the system apart, study its parts in relative isolation, and then use what we learn about the parts in this way to explain the behaviour within the whole. As Kaiser puts it:

> 'reductive explanations include only information that is discovered for example by taking the parts out of their original whole and studying them for instance in vitro, or by changing the original context for instance by inserting an alien object into the original whole and exploring the parts in this different context'.
>
> (Kaiser, 2015, 229)

The parts-in-isolation character of reductive explanation is very obvious in the case of fundamental physics, where, for example, our understanding of the interaction between otherwise isolated pairs of charged particles is used to explain the behaviour of these same particles when they form part of a larger system. According to Kaiser, explanations that study parts in isolation are also common in biology. She offers as an example the explanation of how the molecule Na^+/K^+-ATPase actively transports Na^+ ions out of the cell and pumps K^+ ions in against their electrochemical gradients (Kaiser, 2015, 232–4). The Na^+/K^+-ATPase 'pump' is a protein that is embedded in the cell membrane. The explanation of its activity involves a number of steps (see Elliott and Elliott, 2005, 115–16 for further details):

1. In its resting state the Na^+/K^+-ATPase pump is able to bind with three Na^+ ions from within the cell.

2. Once the three Na^+ ions are bound, the pump can catalyse a reaction (hydrolysis) which splits an ATP molecule into ADP plus a phosphoryl group. The phosphoryl group attaches to the pump.

3. The binding of the phosphoryl group triggers a conformational change in the pump molecule (i.e. a change in shape) such that it opens to the space outside the cell and loses its affinity to bind with Na^+ and so these ions are released outside the cell. At the same time the pump gains an

affinity to bind with K^+, and thus binds with two K^+ ions from outside the cell.

4. Once the two K^+ ions are bound, the pump loses its affinity to bind the phosphoryl group, and the group is released. This triggers a return to the original conformation, in which the pump is open to the inside of the cell and has an affinity with Na^+ ions rather than Ka^+ ions. Thus the Ka^+ ions are released inside the cell and the process returns to step 1.

This process involves a sequence of interactions between ATP, Na^+, K^+ and various subunits of the Na^+/K^+-ATPase molecule (i.e. the binding sites). Kaiser says that these interactions can be studied in isolation because the relevant properties of the parts of the Na^+/K^+-ATPase are 'intrinsically determined'. For example, she says:

> it can be argued that the dispositional property of a particular subunit to bind Na^+ or K^+ (when it is present) is due to the properties of the amino acid residues it is composed of. Accordingly, the properties of the biological parts of the Na^+/K^+-ATPase can be studied in isolation from their original context (e.g. in isolation from other parts of the Na^+/K^+-ATPase).   (Kaiser, 2015, 234)

Now molecular biology is only a couple of steps away from physics—the 'binding' of one molecule to another involves electromagnetic forces of attraction between the two molecules, and the shape of a protein (which itself is determined by the balance of electromagnetic forces between its parts) can affect the protein's ability to bind other molecules by exposing or hiding parts that have an appropriate pattern of electric charge.[1] Thus, it is not surprising that explanations in molecular biology will often share the parts-in-isolation character of explanations in physics. But the study of parts in isolation is a feature of higher-level biological explanations as well.

Consider explanations in the field of population ecology, which investigates the dynamics of species populations over time and space. It is common for explanations in population ecology to use models that treat individual species as separate parts of the ecological system, and then investigate the interactions between species one at a time. If further detail is needed, the focus might shift to the level of individual organisms, but once again these interactions are considered at first in isolation. Consider, for example, the following discussion of predator–prey interaction taken from a popular textbook in ecology:

> What roles do predators play in driving the dynamics of their prey, or prey play in driving the dynamics of their predators? Are there common patterns of dynamics

---

[1] Indeed, the last couple of decades have seen a lot of success in using methods from physics to construct computational models of molecular processes like this. See, for example B. Corry (2016).

that emerge? The preceding sections should have made it plain that there are no simple answers to these questions. It depends on the detail of the behaviour of individual predators and prey, on possible compensatory responses at individual and population levels, and so on. Rather than despair at the complexity of it all, however, we can build an understanding of these dynamics by starting simply and then adding additional features one by one to construct a more realistic picture.

We begin by consciously oversimplifying—ignoring everything but the predator and the prey, and asking what underlying tendency there might be in the dynamics of their interaction. It turns out that the underlying tendency is to exhibit coupled oscillations—cycles—in abundance. With this established, we can turn to the many other important factors that might modify or override this underlying tendency. Rather than explore each and every one of them, however, Sections 7.5.4 and 7.5.5 examine just two of the more important ones: crowding and spatial patchiness. These two factors cannot, of course, tell the whole story; but they illustrate how the differences in predator-prey dynamics, from example to example, might be explained by the varying influences of the different factors with a potential impact on those dynamics.   (Townsend et al., 2008, 233)

The authors explicitly state that interaction between predator and prey species is first considered in isolation. Other interactions are then added to build up a more complex, and more realistic model.

Note that the 'taking apart' that is characteristic of reductive explanation may be something quite literal, as when a molecular biologist extracts a protein from a cell and places it in a test tube to study the reactions that the protein catalyses. The 'taking apart' could also be merely conceptual, as when theoretical physicists posit the existence of quarks and theorize that the properties of protons and neutrons can be explained by the properties of, and interactions between, the quarks that make them up (these quarks are never literally studied in isolation, since one of the features of quarks is that they can never actually be isolated—pulling a bound pair of quarks apart requires a lot of energy, and this energy instantly creates two more quarks, each bound to one of the original pair).

Note also that talk of studying the parts 'in isolation' is somewhat misleading, since in practice nothing can ever be truly isolated. The protein in the test tube, for example may be influenced by contaminants in the water in which it is suspended, the electromagnetic field generated by the lighting in the laboratory, vibrations caused by the footsteps of the lab technicians, and so on. It is only in theory that we can truly isolate the parts, and even then we rarely consider parts in true isolation: studies of the interactions between Na^+ ions and Na^+/K^+-ATPase, for example, often assume that the Na^+ ions are dissolved in water, and studies of interactions between a predator and a prey species always assume a background environment capable of sustaining life. For simplicity's sake, I will continue to speak of studying parts in isolation, but, as Kaiser notes and, as we shall see in Chapter 3, what is really crucial is that the parts are studied in one context (usually a simplified one) and what we learn of the parts in this context is used to explain the behaviour of a system in which these parts are in some other context.

As it stands, (3) seems to have a problem. Reductive explanations are typically causal explanations, explaining the behaviour of the whole in terms of causal interactions between its parts (as is noted in the parenthetical gloss of (3)). But a part in isolation cannot interact with anything; interactions require the presence of more than one object. Consider our reductive explanation for the behaviour of falling water. This explanation refers to the attraction between water molecules and we could not learn about this attraction by studying a single water molecule in isolation. In order to study this attraction, we need to study (either experimentally or theoretically) *pairs* of water molecules.[2] Thus, if we are limited to studying the parts in isolation, then it seems we will not learn anything about their interactions, and so cannot refer to such interactions in a reductive explanation.

In fact, however, the problem is merely apparent. Suppose that W and X are water molecules that are parts of some body of water Y, then the pair (W, X) taken together will also be a part of Y.[3] So, we can learn about the interaction between W and X by studying in isolation the part that is constituted by both W and X. Note that when I speak of the pair (W, X) taken together, I am not speaking of the set whose elements are W and X, for sets cannot be parts of physical objects (at least as sets are usually understood). Rather, I am speaking of the *physical system* that is constituted by W and X. Importantly, this physical system will include physical relations between W and X, such as the distance between them. With these thoughts in mind, I believe that we can clarify our characterization of reductive explanations by introducing the concepts of basic objects and subsystems as follows:

Definition 1 (Basic Objects). An object is *basic* with respect to an attempt at reductive explanation iff the explanation treats these objects as having no parts.

Basic objects are not necessarily fundamental atoms, they may themselves be complex entities like planets, people, or breeding pairs. The point is that they are *treated* as atomic by the explanation—their internal structure is not

---

[2] One might object that we could learn about the attraction between water molecules by studying the electromagnetic field generated by a single water molecule, and studying how single water molecules interact with electromagnetic fields. But this approach posits a new entity, the electromagnetic field, and studies the interactions between the pair that consists of the water molecule and the field.

[3] Here I have made two minimal mereological assumptions. The first is that the mereological sum of W and X exists. This seems reasonably uncontroversial, since we have already assumed that W and X form part of a larger object which is the body of water Y. Thus, we have effectively assumed that none of the standard worries about unrestricted composition applies in this case. Second, I have assumed that if W is a part of Y and X is a part of Y, then the mereological sum of W and X is a part of Y also. This assumption follows directly from the concept of mereological sum that is found in, for example, Eberle (1967), Bostock (1979), and van Benthem (1991). These two assumptions are built into the definitions of *system* and *subsystem* below.

relevant. This is why the definition of a basic object is relative to an attempt at explanation. I will consider the concept of a basic object in more detail in Chapter 5.

Definition 2 (Systems). A *system* is a collection of basic objects together with all their properties and relations to each other.

Definition 3 (Subsystems). A system X is a *subsystem* of system Y iff X is constituted by a subset of the basic objects that constitute Y.

We are now in a position to state clearly what I take to be the essential features of a reductive explanation. I take this to be a generalization and clarification of Kaiser's characterization.

Definition 4 (Reductive Explanation). A reductive explanation explains some behaviour within a complex system Y by:

(1) representing Y as composed from multiple basic objects and referring to the properties of, relations between, and interactions between these objects, and
(2) representing the subsystems of Y as if they were systems in isolation (that is, representing only those relational properties of, and interactions within, subsystems that can be studied in other contexts than *in situ*).

Note that this characterization of reductive explanation is not limited to reductive explanation in biology. Indeed, it is not even limited to physical systems. If there are non-physical souls, for example, then our characterization of reductive explanation allows the possibility that we could give a reductive explanation for the behaviour of such souls, so long as those souls have parts. Note also that the basic objects of the reductive explanation need only be basic with respect to that explanation; there is no requirement that these basic objects be fundamental in any broader sense. It is possible to provide reductive explanations of emission spectra in terms of interactions between electrons, photons, and atomic nuclei; of ecosystem behaviour in terms of interactions between organisms; and of car engines in terms of carburettors, pistons, spark plugs, and the like. What is more, the basic objects need not all belong at the same level of organization. As Kaiser (2015, 190–2) points out, for example, one can give a reductive explanation of photosynthesis that refers to photons and electrons—which seem to lie at a fundamental level of organization—as well as complex molecules like chlorophyll, ATP, and ATP Synthase, which all seem to lie at a higher level of organization.

Reductive explanation, as I have described it here, explains behaviours of a complex system in terms of the properties of, and interactions between, the system's parts. Recently, however, Carl Gillett (2016) has developed an excellent account of scientific reduction which places much more emphasis on *processes*. Inspired by Gillett, one might argue that what really goes on in scientific reductions is that processes at one level are shown to be composed of processes at a lower level, and thus the high-level processes are explained in terms of their component processes, rather than individual objects. I have no objection to the language of processes, for, like Gillett, I take the relevant kind of process to consist in 'a power of some individual being triggered and manifested to produce some effect' (Gillett, 2016, 64). Thus, when we explain a process at the high level in terms of lower-level processes, we are, *ipso facto*, explaining the high-level process in terms of the triggering of powers of individuals at the lower level. That is to say, when we explain high-level processes in terms of component processes, we are explaining these high-level processes in terms of the properties of, and interactions between, the system's parts.

2.3 Mechanistic Explanation

In the past few decades there has been much interest among philosophers of science, and particularly among philosophers of biology, in *mechanistic explanation* (Glennan, 1996; Machamer et al., 2000; Darden, 2002; Glennan, 2002; Woodward, 2002; Bechtel and Abrahamsen, 2005; Baetu, 2015; Glennan, 2017). This literature inherits the term 'mechanism' from seventeenth-century attempts to explain phenomena as diverse as the motion of the planets, the properties of light, and the flow of blood around the body in terms of parts that have only geometric properties and that interact by bumping into each other, thereby pushing each other around. Descartes's *Le Monde* is the classic example of this early mechanistic approach to scientific explanation. The 'new mechanists', however, have a much broader view of mechanism. In order to accommodate the kinds of explanation that have appeared in the sciences since the seventeenth century, we need to allow mechanisms whose parts have non-geometric properties (such as electric charge or mass), and whose interactions can involve more than pushing each other around (they can also, for example, attract, repel, bind, energize, diffuse, or conduct). Nonetheless, the recent literature retains the idea that in a mechanistic explanation one attempts to explain the behaviour of a complex system by looking for the mechanisms that bring that behaviour about. There is some disagreement over the details of what counts as a mechanism, but there is a common core to all modern views, which Stuart Glennan puts as follows: 'A mechanism for a phenomenon consists of entities (or parts) whose activities and interactions are organized in such a way that they produce the phenomenon'. (Glennan, 2017, 17).

Clearly, there is much similarity between mechanistic explanation and reductive explanation, and there will be many explanations that count as both. However, the two types of explanation are not the same, since there may be mechanistic explanations that do not count as reductive explanations. As Kaiser (2015, 239–41) notes, the core concept of mechanistic explanation does not include the requirement that subsystems be treated as systems in isolation, and many of the new mechanists claim that mechanistic explanations often refer to whole-of-system organizational properties that can only be studied *in situ*.

William Bechtel and Robert Richardson (2010), for example, suggest that, in systems that involve positive or negative feedback loops, the organization of the system as a whole can be more important than the distinctive contributions of its constituent components, and so, they say, the system cannot be understood via decomposition into a set of independent subsystems. I am not convinced that this kind of non-linearity is incompatible with the reductive method of explanation, but I will not argue this point here. There is, however, another way in which mechanistic explanations may violate the systems-in-isolation character of reductive explanation. The components of a complex system often behave quite differently *in situ* than they would in isolation: a pair of water molecules in isolation will tend to stick together, while that same pair may be torn apart when accelerated through the air; a particular sequence of DNA may do nothing of interest in isolation, but may cause cancer when it appears in a particular place in the genome of a living organism. This difference between behaviour in isolation and behaviour *in situ* poses a prima facie problem for the reductive method, which I will address in the next chapter. Mechanistic explanations, however, often sidestep this problem by focusing only on the behaviours exhibited by component mechanisms *in situ*, or in very similar contexts. If it is known, for example, that a particular DNA sequence leads to the production of a particular protein when this sequence appears at a particular location in a particular chromosome in a particular type of living cell, then mechanistic explanations can refer to this fact without worrying about what this particular sequence of DNA would do in some other context. Such explanations do not respect the systems-in-isolation approach, and so will not be reductive in the sense I have defined here (though I note that a reductive explanation may be needed if we want to explain *why* that particular sequence of DNA leads to the production of that particular protein in that particular context).

2.4 Conclusion

The point of this chapter was to make clear exactly what I have in mind when I speak of reductive explanation. Reductive explanations attempt to explain the behaviour of complex systems by reference to the interactions of basic component objects. In particular, the kind of explanation I am interested in proceeds by breaking the complex system up (physically or conceptually) into smaller subsystems and studying the interactions between the basic objects within these subsystems. Our knowledge of the interactions within subsystems is then brought together to explain the behaviour of the relevant part of the complex system.

This method of explanation is common in physics and, as Paul Davies (2006, 35) has suggested, it is particularly successful in fundamental particle physics. The concept of reductive explanation is very similar to, but not identical to, the concept of mechanistic explanation that is currently a hot topic in the philosophy of biology. Nonetheless, as Kaiser notes, reductive explanations are common in biology, and many mechanistic explanations in biology will count as reductive explanations. Reductive explanations, I believe, are also common in medical sciences, social sciences, economics, and indeed in everyday non-scientific contexts (for example, when we explain the change in temperature of the room in terms of interactions between heaters, windows, open doors, and the air in the room—see Chapter 4).

The ubiquity of the reductive method of explanation suggests that it is a topic well worth studying. Furthermore, if, as I suggest in the following chapters, the reductive method makes various assumptions about the metaphysics of properties and of causation, then the success of the method in so many different realms would seem to have important implications for our understanding of the deep structure of reality.

3
Causal Influence

When applying the reductive method of explanation, as I have defined it in Chapter 2, we use our knowledge of, or hypotheses about, how the subsystems of a complex system behave individually to make inferences about how they will behave together. If this strategy is to work, then our knowledge of the subsystems as individuals must continue to apply in some sense when these individuals are part of the larger system. But this epistemic assumption only seems justified if a corresponding metaphysical assumption is true. That is, the reductive method presupposes that something about the component subsystems remains constant when these components are put together in the larger system.[1]

This presupposition of the reductive method has been noted by others and is often the target of critiques of reductionism. Marc Van Regenmortel, for instance, tells us that 'reductionists analyse a larger system by breaking it down into pieces and determining the connections between the parts. They assume that isolated molecules and their structure have sufficient explanatory power to provide an understanding of the whole system' (Van Regenmortel, 2004, 1016). William Wimsatt calls the assumption *Articulation-of-Parts Coherence*, which he describes as:

> Assuming that the results of studies done with parts studied under different (and often inconsistent) conditions are *context-independent*, and thus still valid when put together to give an explanation of the behavior of the whole. This can apply not only to intra-organismal parts and systems but at other levels—e.g. to the combination of behavior of individual species to analyze the behavior of an ecological community. (Wimsatt and Sarkar, 2006, 470)

Both Wimsatt and van Regenmortel then go on to suggest that this assumption is unwarranted. Richard Levins and Richard Lewontin put this concern more bluntly. The method of isolating parts as completely as possible and studying these isolated parts, they say, is the result of the 'reductionist myth of simplicity' (Levins and Lewontin, 1980, 76).

---

[1] I am only claiming that the existence of such invariants is necessary for the success of the reductive method, not that it is sufficient. In addition, we will need laws that tell us how these invariants combine in complex systems. I discuss laws of composition in Sections 3.6 and 5.4 below.

These authors are right to point out the role that this assumption plays in the reductive method of explanation, and they are right to note that we have no independent reason to believe that the assumption will always hold, and hence that there may be contexts in which the reductive method will not be successful. However, these authors take a rather pessimistic view of this situation, complaining, as we might put it, that the glass of reductionism is half-empty. I prefer to view the glass of reductionism as half-full: it is an empirical fact that the reductive method of explanation has been remarkably successful in many contexts. This empirical fact gives us good reason to believe that the assumptions of the reductive method are satisfied within many contexts.

In particular, the remarkable success of the reductive method gives us good reason to believe that in many contexts, component subsystems possess an invariant feature which satisfies the following three conditions:

1. The feature is invariant across the relevant contexts (or at least invariant enough for practical purposes).

2. The feature is something we could (at least in principle) learn about whilst studying a subsystem in relative isolation.

3. The knowledge gained by studying the feature in the simple context must be useful for understanding behaviour in the complex context.

I will argue below that, in the interesting cases at least, the invariant feature must involve a new kind of entity that I call *causal influence*. Causal influences are not explicitly included in most standard ontologies, though something like them has been discussed by George Molnar (2003), Nancy Cartwright (2009), and Stephen Mumford and Rani Anjum (2011a). Thus, consideration of the reductive method provides a strong argument for a relatively novel ontology. What is more, since the reductive method is most spectacularly successful in the context of fundamental physics, it is likely that this novel ontology is also fundamental; that is to say, there is good reason to think that causal influences are a fundamental part of reality. Not only is the glass of reductionism half-full, but the liquid it contains is particularly nourishing for a metaphysician of science.

My argument for the existence of causal influence proceeds by considering, and discarding, other ontological types which might conceivably fill the role of the invariant feature presupposed by the reductive method. After eliminating all the other possibilities that seem relevant, I posit causal influence as a type of entity that has just the required features to play this role. Without further ado, then, let us consider the possible candidates in turn.

3.1 Invariant Behaviour

Let us begin our investigation by focusing on the options that are available in a Humean metaphysics. The Humean metaphysical world view is a popular one in contemporary philosophy, with David Lewis, Barry Loewer, and Helen Beebee among its notable adherents. The defining claim of the Humean metaphysics is that there are no 'necessary connections' between distinct entities, so that the possession of one property does not necessitate anything regarding the possession of any other distinct property. In particular, the Humean metaphysics does not include any properties that are essentially dispositional in nature. As will be discussed in greater detail in Chapter 4, dispositional properties typically refer to other properties (fragility, for example is a disposition to break when struck, and so is intimately connected to the properties of being struck and being broken), and so the Humean holds that dispositional properties must all be reducible to more fundamental non-dispositional properties.

Turning our attention back to the invariant that could ground the reductive method, perhaps the simplest view is that it is the behaviour of the components that remains invariant between contexts. What I have in mind by the term 'behaviour' is simply the temporal sequence of properties (including relational properties) the system actually possesses. So, to say that a particular behaviour is invariant accross contexts is to say that a particular sequence of properties occurs in all these contexts. Our focus on a Humean metaphyics means that the properties in question are properties such as position, distance, size, velocity, mass, charge, and spin, as well as macroscopic properties that may supervene on these, like shape, colour, and temperature.

The suggestion that it is simply behaviour that is invariant between relevant contexts has two things going for it: (1) We learn about systems (whether in complex or simple contexts) by observing their behaviour; (2) it is the behaviour of systems that we wish to predict and/or explain by employing the reductive method. Thus, if behaviour were invariant between the relevant contexts, then it could certainly serve the purposes of the invariant that is presupposed by the reductive method.

There are, indeed, numerous cases in which the relevant behaviour of a system remains invariant across relevant contexts. Consider, for example, a system consisting of two intermeshed gears. As one gear turns clockwise, the other will turn anticlockwise, and the ratio of their angular velocities will be equal to the ratio of the number of teeth on each gear. These facts about the behaviour of the two-gear system are invariant throughout a large range of contexts and so can be referred to by reductive explanations in these contexts. In this way, for example, we can use our knowledge of the behaviour of subsystems consisting of two intermeshed gears to explain the behaviour of complex systems like mechanical clocks (in a simple case, we may have one subsystem consisting of the drive gear and the gear attached to the minute hand, and a second subsystem consisting of the same drive gear and the gear attached to the hour hand).

Of course, a system of two intermeshed gears will not behave in the way described above in *all* contexts. If we were to add a third gear that meshes with both the gears in our original two-gear system, then the system will lock up, and the two original gears will no longer behave as described. The two-gear system may also diverge from the expected behaviour if we were to cover the gears in glue, or introduce grit into the system, or hit them with a hammer. We can only use our knowledge of the behaviour of a two-gear system to explain the behaviour of a complex system in contexts in which the two-gear system is not interfered with. Clocks and other clockwork machines are carefully designed to allow their two-gear subsystems to behave as expected, and the mechanisms are generally encased in a protective shell to reduce the chance of interference from outside. It is for this reason that we can successfully use our knowledge of the behaviour of two-gear systems to explain the behaviour of a large variety of clockwork machines.

Mechanistic explanations in biology also often rely on invariance in the behaviour of simple subsystems. Indeed, in their landmark paper on mechanistic explanation in biology, Peter Machamer, Lindley Darden, and Carl Craver define mechanisms in a way that suggests they have invariant behaviours in mind: 'Mechanisms are entities and activities organized such that they are *productive of regular changes* from start or set-up to finish or termination conditions'(Machamer et al., 2000, 3, my emphasis). Consider, for example, the classic mechanistic expla-nation of the transmission of a nerve signal from one neuron to another across the synaptic gap that separates them. As Machamer, Darden, and Craver put it, 'a presynaptic neuron transmits a signal to a post-synaptic neuron by releasing neurotransmitter molecules that diffuse across the synaptic cleft, bind to receptors, and so depolarize the post-synaptic cell' (2000, 3). This explanation is intended to apply to transmission of nerve signals in many parts of the body, under many different circumstances, and, indeed, in many different species. Thus, it is assumed that the described chain of behaviours remains sufficiently invariant across all these contexts.

Like clockwork mechanisms, biological mechanisms have been (naturally) selected for their ability to exhibit regular behaviours, and they typically take place in an environment that offers some protection from outside interference. In the terminology of Nancy Cartwright (1999, ch 3), artificial and biological mechanisms often occur as parts of a *nomological machine* which allows them to exhibit consistent and regular behaviour.

Reductive explanations, however, are not limited to such nomological machines. Consider, for example, our burgeoning understanding of the role that genetics can play in the development of numerous diseases. In most cases where a particular gene is implicated in the development of some particular disease, it is not the case that possession of that gene invariably results in suffering from that disease. The outcome at the level of the entire organism may depend not only on that gene but on other genes that the organism possesses, and the environment in which the organism develops. Nonetheless, reductive explanations for a disease may still be given in terms of the contribution that a particular gene makes to the development of the disease (as compared, say, with the contributions made by different features of the environment).

Reductive explanations outside biology and engineering typically do not involve nomological machines and so cannot rely on invariance in the behaviour of subsystems. This is clearest in the case of reductive explanation in physics. Consider the explanation for the behaviour of falling water mentioned in Chapter 1. This explanation appealed to the attraction that water molecules feel for each other. This attraction is felt between any pair of water molecules, and so the relevant subsystems here are pairs of such molecules. Now, in isolation a pair of water molecules placed in close proximity will move towards each other and stick together. However, there are many contexts in which a pair of water molecules will not behave this way. If they are heated sufficiently, for example, they will fly apart. What is more, the context under consideration is precisely one in which water molecules do not continue to stick together, for we are interested in the way the sheet of water spontaneously comes apart. Nonetheless, it is possible to give a complete reductive explanation of the phenomenon in question, and this reductive explanation refers to what we know about two-particle subsystems. Thus, simple behaviour cannot be the invariant that grounds our reductive explanation of the behaviour of falling water.

In what follows, I will put aside reductive explanations that occur in the context of a nomological machine and focus instead on the more interesting cases in which we are able to provide reductive explanations even though the behaviour of the relevant subsystems is not invariant across the contexts being considered. The crucial point is that we can and do successfully apply the reductive method in many such cases. Since the behaviour of the subsystems is not invariant in these cases, mere behaviour cannot be the invariant that grounds the successful application of the reductive method in these interesting cases.[2] So let us consider what else might fill the role of the invariant that grounds reductive explanation in such cases.

3.2 Invariant Laws

Laws of nature, perhaps, are a more likely candidate for the invariant that grounds the reductive method, for laws are typically taken to be invariant, and they are often thought to be explanatory. Elsewhere in this book, I take the term 'law of nature' to refer to a particular kind of universal statement (see Chapter 7 for details), but the term is sometimes used to refer to some entity or set of facts that might serve as the truth maker for these statements. Whichever way we understand the term, however, laws of nature cannot be the invariant that grounds the reductive method.

---

[2] J. S Mill (1843, Book III, ch. VI) seems to disagree, asserting that, in all cases where the reductive method works, the behaviour of a complex system simply consists in all of the components doing what they would have done in isolation. Mill's position will be discussed in Section 6.2.1.

Consider the view that a law of nature is not just a statement of a particular kind. According to Stephen Mumford (2004), for example, the usual view outside philosophy is that laws of nature are independent entities that somehow govern the behaviour of objects in such a way as to give rise to the regularities that we observe. Let us call such laws 'governing laws'. Mumford argues persuasively that there are no such governing laws, but if he were wrong about this, then one might think that governing laws are just the thing to serve as the invariant that grounds the reductive method.

But consider for a moment what we can learn about governing laws. By hypothesis, governing laws are independent entities, over and above the objects whose behaviour we observe. Thus, a governing law is not something that we can observe directly, and our knowledge of such laws is restricted to knowledge of the behaviours that these laws bring about. Even if governing laws exist, therefore, our knowledge of such laws is restricted to knowledge of the truth of certain *law statements*, which describe the kinds of behaviours to which governing laws would give rise. Regardless of which view of laws we have, then, a reductive explanation which relies on invariant laws must make use of invariant law statements, and, as we shall see, such statements cannot do the job.

Within philosophy of science, law statements are typically taken to be expressible as universal conditionals of the form 'if a certain set of conditions, C, is realized, then another specified set of conditions, E, is realized as well' (Hempel and Oppenheim, 1948, 153). Although law statements in science are not usually given explicitly in this way, the conditional form reflects the fact that our knowledge of almost all laws of nature—whether from physics, biology, economics, or wherever—tells us what will happen only if we supply a set of initial (or final, or intermediate) conditions.

But the conditional form of such law statements leads to trouble for the reductive method. For suppose that we study a system in isolation and learn that it obeys a certain law. That is, we learn that under conditions C (in which the system is isolated) behaviour E always occurs. This knowledge can be of no use to us when we want to know what the system will do as part of a complex—that is, when we want to know what will happen when conditions C no longer obtain. The problem, then, is that although the laws we might discover by studying a system in isolation may be invariant, what we learn about the laws in this way is of no use in predicting or explaining how the system will behave in a more complex situation.

One might object to my assumption in the previous paragraph that conditions C (isolation) will no longer obtain when the system is part of a complex. For it is possible to specify a set of conditions C that will obtain both when the system is in isolation and when it is part of a complex (as a trivial example, let C be the condition that $2 + 2 = 4$). If a law statement refers only to such conditions, then it will continue to apply when the system is part of a compound. In response, I accept that there may be such law statements, but let us follow through their consequences. Suppose we learn that 'if a certain set of conditions, C, is realized, then another specified set of conditions, E, is realized as well', and suppose also that C is realized both when the system is in isolation and when it is part of a compound. Then it follows that E will be realized both when the system is in isolation and when it is part of a compound. That is, the relevant behaviour of the system will not change from one context to the other. But I have already considered situations in which behaviour remains invariant, and have set such situations aside as uninteresting.[3] In interesting situations, the behaviour of our component system will not be invariant, and hence the relevant conditions C cannot be realized both when the system is in isolation and when it is part of a compound.

Perhaps a more sophisticated view of laws will help. Consider Newton's Second Law: $f = ma$. This law statement is not equivalent to a universal conditional of the sort suggested by Hempel and Oppenheim; rather, it is equivalent to an infinite number of such conditionals, one for each possible value of f and m. Thus, if we can learn that a system obeys this law by studying it under one set of conditions (i.e. particular values for f and m), we might still be able to apply the law when the system is subject to a different set of conditions.

Unfortunately, the move to a more sophisticated view of laws does not really help. The same problem simply raises its head in another guise. The problem we face is one that has been pointed out on numerous occasions by Nancy Cartwright (see, for example, Cartwright 1980; 1983; 1989; 1999). Cartwright observes that the standard view of law statements is that they describe regularities in behaviour. But, she says, such regularities are exceedingly rare in nature; the actual behaviour of a system depends on the environment it is in, and there will often be external influences that disrupt any law-like regularity. Thus, law statements—conceived of as statements of regularity in behaviour—will most likely be false.

So, for example, taking a sophisticated view of the laws of classical mechanics, we might interpret these laws as implying that massive bodies will accelerate towards each other at a rate proportional to the product of their masses and inversely proportional to the distance between them. Thankfully, however, I can report that the computer I am typing on at this moment is doing no such thing. The computer is not accelerating towards the Earth because it is supported by a desk. The desk seems to have destroyed the supposed regularity.

Relating this example back to the reductive method, the problem is this: if I study two massive bodies in isolation, I will find that they obey a law according

---

[3] Perhaps I am being a little unfair labelling these situations 'uninteresting'. As Cartwright (2009, 154–5) points out, complex systems modelled by a system of simultaneous equations might be interpreted in such a way that each equation specifies the characteristic behaviour of a component system. Since a solution to a system of simultaneous equations is one in which each individual equation is satisfied, one might argue that such solutions describe cases where the characteristic behaviours of the components all occur simultaneously. I do not really mean to suggest that all such systems are uninteresting. What I really mean is that they are not the systems I am interested in here.

to which the bodies will accelerate towards each other at a rate proportional to the product of their masses and inversely proportional to the distance between them. However, when I place these two bodies into a more complex system involving non-gravitational forces (such as electromagnetic forces in the case of the desk), they will no longer accord with this law statement. Thus, the law is of no use to the reductive method (at least this is true if we conceive of the law statement as describing a regularity in the behaviour of a system as conceived above—I will make use of the ontology developed below to consider other ways to conceive of law statements in Chapter 7).

Before giving up on laws of nature, let us consider an objection that Cartwright briefly considers and rejects. This objection suggests that what we need to do is employ a more complex 'super law' which takes into account all the relevant interactions into which a system can enter. In the example above, the relevant super law statement would be one 'which says exactly what happens when a system has both mass and charge' (Cartwright, 1980, 81). There are two problems with this suggestion. The first problem is that the way we actually come up with such a super-law statement is by studying the interactions involved in isolation. Once we know how the individual interactions work, we put this knowledge together to formulate a super-law statement. So our knowledge of the super law is a result of an application of the reductive method, and so cannot be something we rely on in that very application. The analysis that produces the super-law statement must rely on an invariant other than the super law itself.

The second problem with the super law suggestion is that, as Cartwright points out, 'if we fail to describe the component processes that go together to produce a phenomenon, we lose a central and important part of our understanding of what makes things happen' (1980, 81). The reason we miss this important part of our understanding is that super-law explanations are simply not examples of the reductive method.

3.3 Invariant Humean Dispositions

We have seen that neither behaviour nor laws describing behaviours can play the role of the invariant that grounds interesting applications of the reductive method. The next obvious candidate to consider is that it is the dispositional properties of component systems that play this role. The thought would be that we observe the components of a system individually under various different circumstances and thereby learn about their dispositions to behave. We then apply our understanding of these dispositions to explain the behaviour of the components when they interact to form a larger system, and thus explain the behaviour of the complex whole.

Recall that, within the Humean metaphysics, dispositional properties must ultimately be reducible to non-dispositional properties. The first step in such a reduction is typically an analysis of dispositions in terms of subjunctive conditionals something like the following: 'X is disposed to behave in manner B under conditions C iff were X to be placed in conditions C, it would behave in manner B'. Indeed, such an analysis is an almost universal starting point even within non-Humean ontologies (as in Chapter 4). Although this straightforward conditional analysis has been elaborately modified to avoid certain objections (see, e.g., Prior, 1985; Lewis, 1997; Malzkorn, 2000; Mellor, 2000), the basic strategy is still to analyse dispositions in terms of the behaviour that would occur if certain conditions held.

But if the behaviour mentioned is the kind of behaviour that is possible in the Humean metaphysics, then we will run into exactly the same problem that faced laws of nature. When we observe a component in isolation, we learn about its dispositions to behave under certain conditions C. But to explain the behaviour of a compound system, we need to know how the components are disposed to behave under very different conditions C', conditions in which the component is part of the larger system. Furthermore, we have seen that in any interesting case, the actual behaviour of the components in the system will differ from their behaviour in isolation. Thus, the differences between conditions C and C' are clearly relevant to the dispositions in question. In short, the problem is that any behavioural dispositions we discover when a component is in isolation will be dispositions whose antecedent is not satisfied when the component is part of the larger system. Thus, although the dispositions of the components may remain invariant from one situation to another, the dispositions we learn about when studying the components in isolation will not be the dispositions that are manifested when the components are placed into a complex. Simple dispositions, as standardly understood, cannot be the invariant that grounds the reductive method.[4]

Just as there is a sophisticated view of laws of nature according to which a law corresponds to an infinite set of conditionals, there is a sophisticated view of dispositions according to which a disposition corresponds to an infinite number of conditionals. So, for example, we might say that when affected by a force of f Newtons, an electron is disposed to accelerate at a rate of $9 \times 10^{-31} f$ metres per second per second. Complex dispositions like these could be regarded as a kind of multi-track disposition (see Ellis and Lierse, 1994, 29; Bird, 2007, 22).

---

[4] This argument could be taken as an argument against the standard Humean conception of dispositions rather than just as an argument that such dispositions could play the role of the invariant that grounds reductive explanation. However, a defender of the standard concept might respond that an ontology of standard dispositions is required for reasons other than for reductive explanation.

Ultimately, I think that complex dispositions of this sort are a step in the right direction. However, we are not out of the woods yet. For, so long as we analyse dispositions in terms of conditionals which state what behaviour will occur when certain conditions are satisfied, each disposition will correspond one-to-one with a law-like regularity, and hence will be subject to the same problems that faced laws of nature. For example, an electron will have a disposition to behave one way in the presence of massive objects, and a disposition to behave another way in the presence of charged objects, and neither of these dispositions will be manifest when the electron is in the presence of an object that is both massive and charged. We might try to solve the problem by positing a 'super disposition', but this will face the same problems as a super law.

3.4 Invariant Capacities or Non-Humean Dispositions

At this stage I believe that we have exhausted the possibilities available to a Humean metaphysics, according to which fundamental reality consists only of objects with categorical (that is, non-dispositional) properties. Laws of nature, on this view, are descriptions of contingent regularities among these objects and properties. Dispositions are descriptions of contingent regularities in the behaviour of the object having the disposition. Thus, laws and dispositions must be analysed in terms of conditionals like those considered above. What I think our discussion so far has shown is that the scientific method assumes an ontology richer than the sparse world described by the Humean.[5]

So what must we add to the Humean ontology in order to ground the reductive method? A number of authors (e.g. Harré and Madden, 1975; Cartwright, 1989; Ellis and Lierse, 1994; Ellis, 2001; Molnar, 2003) have suggested, for various reasons, that the Humean ontology must be augmented with *Tendencies*, *Powers*, *Capacities* and such like. I believe that the addition of such non-Humean concepts is a step in the right direction; however, not just any old non-Humean Tendency, Power, or Capacity will do the job. In order to see what is needed, I think it will be useful to consider the views of Brian Ellis and Caroline Lierse, on the one hand, and Nancy Cartwright, on the other. Let us begin, then, with Ellis and Lierse.

3.4.1 Ellis and Lierse on Dispositions and Powers

It is common to distinguish dispositional properties from non-dispositional or 'categorical' properties (so called since they do no more than categorize their

---

[5] And what O'Connor (2009, 191) brands the 'second-order Humean metaphysics' of David Armstrong (1997) and Michael Tooley (1987) will, I believe, face the same objections.

bearers)[6]. The Humean view described above holds that all dispositional properties supervene on categorical properties together with the laws of nature. This view— called *Categorical Monism* by Mumford (1998) and Bird (2007) and *Categorical Realism* by Ellis and Lierse (1994)—takes the analysis of dispositions in terms of conditionals as a definition. Understood this way, the analysis can provide a path to reduce dispositional properties to non-dispositional properties plus laws of nature. So, for example, on this view to say that a glass is fragile is simply to say that if certain circumstances obtain, the glass will (or is likely to) break, and the reason the glass will (or is likely to) break in such circumstances is that the glass has a certain structure and there are laws of nature that imply that when objects with this structure are placed in these circumstances, they will (or are likely to) break.

Ellis and Lierse deny categorical monism, arguing that some dispositions cannot be reduced to a categorical basis. Their thought is that the conditional analysis is not a *definition* of what it is to have a dispositional property, but, rather, a means of *identifying* such a property. The dispositional property is what makes the conditional true, but it is not simply reducible to the conditional.[7]

With this non-Humean view of dispositions under their belt, Ellis and Lierse define causal powers, capacities, and propensities as special kinds of dispositional property:

> A causal power is a disposition of something to produce forces of a certain kind. Gravitational mass, for example, is a causal power: it is the power of a body to act on other bodies gravitationally. A capacity is a disposition of a kind distinguishable by the kind of consequent event it is able to produce. Thus, for x to have the capacity to do Y is for x to have a disposition to do Y in some possible circumstances. Inertial mass, for example, is a capacity. It is the capacity of a body to resist acceleration by a given force. A propensity is a disposition which a thing may have to act in a certain way in any number of a very broad range of circumstances. For example, the propensity of a radium atom to decay in a certain way in a certain time is a disposition which the atom has in all circumstances.
>
> (Ellis and Lierse, 1994, 40)

These powers, capacities, and propensities are prime candidates for the invariants we have been hunting. For it is at the very least plausible to suppose that these properties of a system remain constant from one situation to another, even though the behaviour they produce might change. Indeed, Ellis and Lierse argue for 'dispositional essentialism', the view that at least some properties have

---

[6] Though note that Mellor (1974; 1982) argues that the distinction is incoherent.
[7] Similar ideas can be found in Harré (1970); Harré and Madden (1973; 1975); Shoemaker (1980); Swoyer (1982); Fales (1990); Fara (2005).

dispositional essences.[8] That is, having certain dispositional properties is part of what makes an entity the kind of thing that it is. If dispositional essentialism is true, then the essential powers, capacities, and propensities of a system must, by necessity, remain invariant as the system is taken from one situation to another. Furthermore, it seems natural to suggest that we can learn about the powers, capacities and propensities of a system by studying that system in simple, controlled circumstances. Finally, Ellis and Lierse have freed these dispositions from the Humean analysis in terms of conditionals, so maybe their view will avoid the problems I outlined above for the standard account of dispositions.

Unfortunately, nothing is quite so simple, for it is not at all clear that Ellis and Lierse's account of dispositions does avoid the problems of the standard Humean account. Although they reject the conditional analysis of dispositions, dispositions are nonetheless only identifiable via conditionals; the disposition to E in circumstances C is simply that property that is responsible for the fact that its bearers would (normally) E in circumstances C. So, suppose that we take a system x and investigate it under a range of simple, controlled conditions C. We find that in circumstances C, x tends to E. Following Ellis and Lierse, we ascribe a dispositional property P to the system x, where P is the property responsible for the fact that in circumstances C, x tends to E. The problem is that this is all we know about P. In particular, we do not know what behaviour P will tend to produce in circumstances other than C.

Consider how this problem plays out in an application of the reductive method. Suppose that I study the way that electrons interact with other electrons by studying pairs of electrons in isolation. I will discover that electrons are disposed to accelerate away from each other. According to Ellis and Lierse, what I have discovered is that electrons have a dispositional property P that is responsible for them accelerating away from other electrons when the two are isolated from other interactions. But what happens if we take the pair of electrons and put them into a more complex system? The answer depends on the details of the system. The two might still accelerate away from each other; they might remain motionless relative to each other; they might even accelerate towards each other. So, can knowing that the electrons have the property P help us predict what will happen? Not so long as dispositions like P are our only addition to the Humean metaphysics. It is true that the electrons' behaviour in the complex system could be the result of the same dispositional property P that led to the two electrons accelerating away from each other in isolation. The problem is that all we learn about P by studying the electron pair in isolation is that possessing P implies a conditional: in situations C (where the

---

[8] This is the now standard understanding of the term 'dispositional essentialism'. Ellise and Lierse understand the term to also include the claim that dispositional properties are determiners of object kinds. This extra condition is not relevant here.

pair is isolated) they will exhibit behaviour E (accelerating away from each other). But we have already seen, in our discussion of Humean dispositions, that such a conditional is no use in predicting what will happen in situations other than C. By interpreting the conditional as a means for identifying the property P rather than an analysis of P, Ellis and Lierse simply shift the problem from metaphysics to epistemology.

3.4.2   Cartwright on Capacities

Cartwright is another supporter of capacities, powers, and tendencies, but she has a slightly different interpretation of the notion from that of Ellis and Lierse. In particular, Cartwright argues that capacities should not be straightforwardly understood in terms of dispositions. Dispositions, she says, are usually tied one-to-one with law-like regularities; they are identified with particular manifestations. Capacities, on the other hand, are 'open ended' and give rise to varied behaviours (1999, 59, 64). If we must think in terms of dispositions, she says, we should think of capacities as being what Gilbert Ryle has called 'highly generic' or 'determinable' dispositions, in the same class as 'being humorous', or 'being a grocer'. Quoting Ryle (1949, 118), she says that such generic dispositions 'signify abilities, tendencies, propensities to do, not things of one unique kind, but things of lots of different kinds' (Cartwright, 1999, 64).

When interpreting Cartwright, we are faced with a dilemma. Either what she has in mind are simply multi-track dispositions of the sort that Ellis and Lierse describe or they are something else. If Cartwright is simply talking about non-Humean multi-track dispositions of the Ellis-Lierse stripe, then surely her view cannot do any better than Ellis and Lierse's in helping us understand the reductive method. If, on the other hand, Cartwright has something different in mind, the question is what?

However Cartwright answers this question, she is going to face difficulties. For the one thing she does tell us is that the behaviour a capacity gives rise to is highly varied. In particular, objects with a given capacity may behave very differently in one circumstance from how they do in another. But now we are faced with the same old problem again: if the behaviour that a given capacity produces can change from one circumstance to another, then how is it that learning about the exercise of a subsystem's capacities in isolation can tell us anything at all about how these capacities are exercised in the differing circumstances of the compound system? Moving from ordinary old dispositions to highly generic dispositions just seems to make our problems worse, for we just add extra variability from one situation to another. The fact that the capacity itself remains invariant is of no use to us if we have no way to know what behaviour the capacity will produce.

3.5 Invariant Causal Influence

The considerations above suggest that none of the obvious possibilities available in popular ontologies (whether Humean or not) is capable of playing the role of the invariant required by the reductive method; we must look beyond standard ontologies. However, it is not too hard to find a model that can make sense of the reductive method. To see this, we simply need to step back and consider the discussion above at a more general level.

The problem identified above is that, in interesting cases at least, the behaviour of a system can change depending on the context it is in. The natural explanation for this is that the behaviour changes due to causal interactions between elements of the system and elements of the environment. The behaviour of an isolated two-particle system, for example, will change if it becomes part of a bigger three-particle system precisely because of causal interactions between the third particle and the two original particles. Similarly, we would expect the dynamics of a population subject to predation by a single species to change if a second predator species is added to the system, and we expect such a change precisely because of new interactions between the prey population and the new predators. Indeed, the reductive method assumes that something like this is going on because the whole point behind the method is to explain the behaviour of the complex system in terms of interactions between its parts. Thus, the problems pointed out above stem from a defining characteristic of the reductive method: it seeks to explain or predict the behaviour of a complex system in terms of the many causal interactions between its parts. The reductive method must proceed by keeping track of and combining these multitudinous interactions.

The considerations above suggest a new model for causal powers, a model that will be investigated throughout the rest of this book. Suppose that a causal power is not a disposition to produce a particular *behavioural outcome* given the right stimulus, but is, rather, a disposition to exert a particular *influence* on the outcome. Suppose, further, that the outcome that occurs is a result of the influences contributed by all of the relevant powers. If the contribution that a power makes remains invariant from one context to another, and if the rules by which the influences compose ('add together') to produce an outcome also remain invariant, then we can use the knowledge of these contributions and rules that we gain in one context to explain and predict behaviour in other contexts. This model of the reductive method identifies not one, but two invariant features. First, the influence that a causal power exerts must remain invariant from one circumstance to another (in a sense to be spelt out below). Second, the laws by which these influences compose must remain invariant.

Suppose, for example, that we are interested in the gravitational behaviour of a system of three objects, *a*, *b*, and *c*. We know that when *a* and *b* are placed nearby in the absence of any outside interference, *b* will exert a gravitational force on *a*, which we can write as f_{ab}. Similarly, we know that when a and c are examined together in isolation, c exerts a force of f_{ac} on a. When a is in the presence of *both* b and c, we assume that b and c will each exert the same force on a that they did when the other was not present. To explain or predict the behaviour of a in the complex system we simply add all the forces acting on it (in this case f_{ab} and f_{ac}) and use the result to tell us how a's motion will change. We can do the same thing with b and c, and so can explain the behaviour of the whole system.

In this example, the forces between pairs of particles are the influences that are assumed to remain invariant, and the resultant behaviour is calculated by composing these forces according to the rules of vector addition (which are also assumed to remain the same regardless of the situation). Forces are paradigm examples of causal influences, and we will spend much time discussing them hereafter. However, it is important to keep in mind that forces are only one example of influence. Forces influence the motion of an object, but it is also possible to influence such diverse things as the temperature of a room, the population of a species, the colour of a flower, and the outcome of an election.

Now, there is a sense in which the force exerted on a by b is not invariant: it varies as the distance between the two objects varies (and the masses of a and b, if these are allowed to vary). What is really assumed is that the force exerted on a by b *when they are a given distance apart* remains invariant regardless of what other objects are about. More generally, when I say that the influence that some object b exerts on some other object a is *invariant*, what I mean is that the influence depends only on the properties of and relations between a and b. Given a particular configuration of the (a, b) system, the influences each exerts on the other will not vary, regardless of what else is going on. That is, the influence depends only on those properties and relations of the (a, b) system that can be studied in isolation. Thus, this invariance is just what is required by the systems-in-isolation character of reductive explanation.[9]

The important point here is that departing from a Humean metaphyscis by adding non-Humean dispositions or powers is not enough: one must also add 'causal influences' that fit between causal powers and the behavioural outcomes that result from these powers. These influences are distinct from the powers themselves, since an object can have a power even when that power is not manifested (and so exerts no influence). Influences are also distinct from the resultant behaviour, since the influence a power exerts is assumed to be invariant from one context to another, whilst the resultant behaviour is not. Thus, our model of the reductive method assumes an ontology with three distinct elements: (1) powers, (2) influences, and (3) outcome states.

---

[9] A slightly more general version of the definition of invariance will be given in Chapter 5.

This threefold distinction is not a part of the standard ontologies mentioned above, but it is not entirely new. George Molnar (2003, 195) noticed the need for such an ontology when he distinguished (1) Powers, (2) Manifestations of powers, and (3) Effects of powers. Cartwright (2009, 151) also makes a similar distinction between (1) The obtaining of a capacity, (2) Its exercise, and (3) The manifest ('occurrent') results. Neither author, however, takes much time to explore the nature of the novel second level of this ontology. What is more, in her previous work Cartwright (1980; 1999) denies the existence of component forces, which seem to be paradigmatic examples of causal influences (see Chapter 6 for further discussion of Cartwright's position on component forces). More recently, Mumford and Anjum (2011a) have explored the contributions of causal powers in some detail; however, as I argue in Chapter 4, they do not pay sufficient attention to the distinction between a power and the influence it contributes. The view I am putting forward here also has similarities to that put forward by Anna Marmadoro (2017), who distinguishes (1) powers in potency, (2) powers in activation, and (3) the activities of powers. Finally, in the quote above Ellis and Lierse distinguish powers, which are dispositions to produce forces, from capacities, which are dispositions to produce a certain kind of event. This distinction suggests that they view the production of a force as distinct from other kinds of event, and so they may recognize the distinction between (2) and (3).

Note that, to support the reductive method, elements of type (2) cannot be reducible to, or analysed away in terms of, elements of types (1) and (3). For, as we have seen, elements of types (1) and (3) alone cannot provide the invariance that is required for the reductive method. So, to accept the existence of elements of type (2) is to make a real addition to one's ontology. This point is recognized by Jennifer McKitrick when she objects to the view as follows:

> On the Molnar/Mumford view, an effect is not called a 'manifestation' because that term is reserved for a newly postulated kind of entity—something that exists in addition to the power and the effect. Lest we multiply entities beyond necessity, we should question whether we have good reasons, empirical or otherwise, to postulate 'manifestations' so understood.   (McKitrick, 2010, 81)

McKitrick is surely correct that we need good reason to add new elements to our ontology; however, we do have good reason in this case. The addition is required in order to account for the applicability of the reductive method.

So what kind of entities are causal influences? In Corry (2006), I argued that influences are vector-like, in that they are characterized by direction and magnitude. This thought is echoed by Mumford and Anjum (2011a), who present what they call a vector-model of causation based on elements of type (2).[10] The idea that influences are vector-like is encouraged by the fact that forces are explicitly represented as vectors. I realize now, however, that things are not as simple as

---

[10] This model will be discussed in more detail in Chapter 4.

I originally thought, so it is worth taking a little time to look at the concepts of direction and magnitude as they apply to influences.

3.5.1 Direction

The reductive method works by studying influences in isolation and using the information so gathered to explain or predict what will go on in more complex situations. Thus, in the first instance, the reductive method characterizes causal influences via the effects that they bring about when acting in isolation. All influences have a characteristic effect (the one it would bring about if acting in isolation), and it is natural to speak of an influence 'pointing towards' its characteristic effect.

It is also a presupposition of the reductive method that the influences produced within a system remain the same whatever situation that system is placed into. So, an influence that points towards effect E in isolation, is still an influence that points towards E when acting in combination with other influences. But, as we have seen, the presence of a causal influence does not guarantee that its characteristic effect will come about, because what actually happens will depend on what other influences are at play. So, an influence continues to 'point to' its characteristic effect even in circumstances where the effect does not come about. The fact that the influence points to this effect, therefore, cannot be analysed away in terms of the effect always following the influence or some such. It is a short step to the conclusion that this 'pointing' is an intrinsic and essential part of an influence being what it is.

A short step, but one that overreaches nonetheless. The reductive method requires only that an influence must have a characteristic effect when acting in isolation, and there must be rules for composing these influences when they act in conjunction. These properties are consistent with the characteristic effect being an essential feature of the influence, but they are also consistent with the relation between influence and characteristic effect being determined by some contingent law of nature. I believe that we do, in fact, have good reason to think that the characteristic effect of an influence is an essential feature of that influence, but nothing we have said so far guarantees this. I will continue to speak of influences as being directed towards an outcome, but until I have argued that influences are essentially directed, this could simply be taken to mean that influences bring about a characteristic outcome when acting in isolation.

3.5.2 Magnitude

There are actually two logically distinct magnitudes associated with each influence. First, there is a magnitude that describes the size of the change that the influence is aiming towards. Compare, for example, the common stinging nettle with the gympie gympie tree of Northern Australia. Both are covered in stinging hairs that can cause pain and irritation upon contact with the skin. The pain produced by the common nettle is relatively mild and does not last very long. The pain produced by the gympie gympie tree, on the other hand, is excruciating and can last for weeks. Intuitively, then, the poison from the gympie gympie tree has a much greater influence on the body than the poison of the common nettle. A simple example can also be found by turning our attention back to forces. Newton's second law, $f = ma$, tells us that the acceleration of an object is directly proportional to the magnitude of the force exerted upon it. Thus, the magnitude of the force is a measure of the size of the change (acceleration in this case) it tends to bring about.

The second magnitude associated with each influence is a measure of the way in which it combines with other influences. The stronger an influence is in this respect, the harder it is for other influences to counteract it. To see that this magnitude is logically distinct from the magnitude that measures the size of the influence's characteristic effect, note that it is conceivable that there could be an influence whose characteristic effect is very small; yet this influence requires a great deal of effort to overcome. Conversely, we can imagine the possibility of an influence whose characteristic effect is very large and yet is very easily overcome.

The distinction between these two concepts of magnitude is easily overlooked, because, as we shall see in Chapter 5, application of the reductive method assumes a systematic relation between the two magnitudes (the algebra of composition of one must be isomorphic to the algebra of composition of the other). In the case of forces, for example, the two magnitudes are effectively identified: the same magnitude is used both when adding component vectors to calculate the total force and when calculating the size of the resultant effect via Newton's second law.

Note also that there is no guarantee that either of the magnitudes associated with an influence can be placed on a ratio scale, in which case they will not have the full structure of a vector space. Forces, of course, can be placed on a ratio scale: one force can be said to have twice the magnitude of another if the characteristic acceleration produced by the first is twice the magnitude of the acceleration produced by the second. Indeed, forces inherit a full vector space structure from the vector space of accelerations via the isomorphism mentioned above (and, in turn, accelerations inherit this structure from the vector space of displacements). Influences directed towards the production of pain, on the other hand, do not seem to have this structure. The pain produced by the gympie gympie tree is greater than that produced by the common nettle, but is it twice as great? One hundred times as great? Intuitively, there does not seem to be a sensible way to answer these questions, which suggests that pain intensity does not have a ratio scale.[11] If it is true that pain intensity does not have a ratio scale, then it would also

---

[11] Though this intuition may be incorrect. Medical researchers and practitioners have developed ratio scales for measuring pain intensity which they believe are valid. See, for example Price et al. (1983).

seem likely that the magnitude of influences directed towards the production of pain would also fail to have a ratio scale.

If some kinds of influence cannot be placed on a ratio scale, then it follows that, in general, we should not take the analogy between influences and vectors too seriously. Thus, rather than describing influences as 'vector-like', I will develop a more generic algebraic account of influences in Chapter 5.

3.6 Composition of Influence

The main point of this chapter has been to track down the invariant that grounds the reductive method in science. My conclusion is that causal influences are what do the job. However, I should probably say a few more words about the second stage of the reductive method—the composition of component causal influences. I have already noted that when the causal influences are forces (that is, they are influences that are directed towards accelerations), we have a clear rule for combining multiple influences: we form the vector sum of the various forces, and the system will behave as if this sum were the only force in play. Similarly precise models can be found in fields as diverse as biology, ecology, and econometrics.

Consider, for example, the explanation of inherited phenotypic traits given in classical genetics. The field of modern genetics is often traced back to Gregor Mendel, who, in 1865, presented his now famous paper 'Versuche über Pflanzen-Hybriden' (Experiments in Plant Hybridization). Mendel created hybrids from various true-breeding strains of pea plant and observed the features of the resulting offspring over multiple generations. For example, he obtained a strain of pea that always had violet flowers and hybridized this with a strain of pea that always had white flowers (by manually pollinating the flowers of one with the pollen of the other). All of the hybrid offspring, he noted, had violet flowers. The violet flower character, said Mendel, is *dominant*, whilst the white flower character is *recessive*. Things got more interesting, however, when he created a second generation by interbreeding members of the hybrid generation. In this second generation white flowers were again present. In particular, he found that three-quarters of the second generation had violet flowers, while one-quarter had white flowers. Mendel studied many characteristics other than flower colour (seed shape, seed colour, albumen colour, pod shape, pod colour, position of flowers, and stem height) and observed this same pattern of dominance by one character in the first hybrid generation followed by a reappearance of the recessive character in later generations, with the ratio of dominant to recessive in the second generation always being around 3:1. What is more, the pattern of inheritance for each characteristic was independent of the other characteristics of the plant. So, for example, the ratio of violet to white flowers in the second generation was 3:1 regardless of seed shape and colour, pod shape and colour, and so on.

Not only did Mendel note these interesting facts, but he was able to explain them with a very simple model of inheritance. Restated in modern terminology, Mendel's model of inheritance runs as follows. Each characteristic (or *trait*, in the modern parlance) of the plant is associated with a 'factor' (or *gene*, as we now call it) that is passed from parent to offspring. So, for example, there is a gene that is associated with violet flowers (call this *A*), and a gene that is associated with white flowers (call this *a*). Furthermore, each pea plant has two genes for any given trait, one that is inherited from the originating egg, and one that is inherited from the originating sperm. If each gene comes in two varieties (*alleles*), *A* and *a*, then there are four combinations (*genotypes*) that are possible for any pea plant with regard to this gene: *AA*, *Aa*, *aA*, and *aa* (where the first letter represents the allele inherited from the egg and the second represents the allele inherited from the sperm). Finally, if a plant has the genotype *AA*, then it will display the trait that is associated with *A* (e.g. violet flowers), if it has the genotype *aa* it will display the trait that is associated with *a* (white flowers), while if it has one gene of each kind, (i.e. *Aa* or *aA*), then it will display the trait associated with the dominant allele (violet flowers). It makes no difference which allele came from the egg and which from the sperm.

With this model, we can explain Mendel's observations regarding flower colour as follows: the original violet-flowered parents all have genotype *AA*. If they are bred with each other their offspring will also be of type *AA* (since they must get one *A* from each parent) and this explains why they are true-breeding. Similarly, the original true-breeding white-flowered parents all have genotype *aa*. The first hybrid generation will all get an *A* from their violet-flowered parent and an *a* from their white-flowered parent and so will all have genotype *Aa* or *aA*. Since *A* is dominant over *a*, all of these peas will have violet flowers. If the hybrid plants are interbred, the second generation that results could get an *A* or an *a* from each parent, and so all four genotypes are possible in this generation. In particular the *aa* genotype is possible and so this generation can contain white-flowering plants. Now, three of the possible genotypes in this generation produce violet flowers (*AA*, *Aa*, and *aA*) and only one produces white flowers (*aa*). So, if we assume that each of the possible genotypes is equally likely, then we should expect that three-quarters of the plants in this generation will have violet flowers and one-quarter will have white flowers. Thus, the observed ratio of 3:1 is explained.

The important point for our purposes is that Mendel's explanation (or at least the modern descendants of his explanation) is a reductive explanation in the sense defined in Chapter 2. The basic objects in the explanation of inherited flower colour, for example, are the flowers, the allele for flower colour inherited from the egg, and the allele for flower colour inherited from the sperm. Each of the alleles exerts an influence on the colour of the flowers, and it is posited that this influence is invariant regardless of the other traits and genes of the plant, so the interaction between allele and flower satisfies the systems-in-isolation character of reductive explanation. Finally the influences of the two alleles are composed to determine the resulting flower colour. Unlike forces, however, the influences of the two alleles do not compose linearly as vectors.[12] But this non-linearity does not mean that we cannot view the alleles as contributing an influence, nor does it mean that we cannot find precise composition laws. Indeed, we can construct a table that characterizes the composition of influences exerted by alleles for flower colour as follows:

	I_V	I_W
I_V	Violet	Violet
I_W	Violet	White

Where I_V is an influence directed towards violet flowers, I_w is an influence directed toward white flowers, and each cell in the table shows the outcome that is produced when the influence indicated on that row is composed with (i.e. acts together with) the influence indicated on that column.[13]

Modern genetics recognizes that the relationship between alleles is not always one of complete dominance; there are also cases of *incomplete dominance* and cases of *codominance* (Urry et al., 2018, 281). These cases can be understood simply as cases in which the composition laws are different. Incomplete dominance occurs when two influences combine to produce an intermediate trait. So, for example, a hybrid snapdragon with one allele coding for red flowers and one coding for white flowers will produce pink flowers. Codominance, on the other hand, occurs when genetic influences combine to produce *both* the traits that the two influences are directed towards. So, for example, a human who has one allele for A blood type, and one allele for B blood type will have AB blood type, which means that her red blood cells will display the antigen characteristic of type A blood *and* the antigen characteristic of type B blood. So the A and B alleles are codominant. Human blood types are interesting because there are three alleles in the population (A, B, and O) and while A and B are co-dominant in relation to each other, they are both dominant over O. Thus, the composition table for the influences of these alleles is as follows:

	I_A	I_B	I_O
I_A	A type	AB type	A type
I_B	AB type	B type	B type
I_O	A type	B type	O type

There are also many cases where we do not have precise rules for the composition for causal influence. Does the lack of precise rules mean that the reductive method

---

[12] Also unlike forces, these influences are not fundamental. The relation between fundamental and composite influences is discussed in Chapter 6.

[13] In Chapter 5 I will suggest that the composition of influences is best seen as producing a *resultant influence*, rather than an outcome. In this case we should regard each cell as indicating the outcome towards which the resultant influence is directed.

cannot be applied? No. Even without precise mathematical models telling us how to compose causal influences we may have rough and ready rules. For example, acids tend to turn litmus paper red; bases turn litmus paper blue. What happens if I place litmus paper into a mixture of an acid and a base? I do not have a precise rule for composing the influences of the acid and the base, but if I know that the acid is strong and the base weak, then I can predict that the litmus paper will tend towards the red end of the spectrum.

As Creary (1981) points out, even when we do not have any idea how to compose the influences at play, the reductive method can still be of use in providing post facto explanations. Suppose that I were to mix an acid and a base together in some unknown proportion. Suppose also that I had no idea about the relative strengths of the acid and the base. In such a situation I would be in no position to predict the outcome of placing litmus paper in the resulting solution. But suppose that I place the litmus paper in the solution and find that it goes red. I can conclude that it went red because the influence of the acid in the solution was stronger than the influence of the base.

However, there must be some restrictions on the form that the laws of composition can take if we are to be able to apply the reductive method. From a practical standpoint, if the laws are too complex, then we will not be able to learn or apply them. But there is a metaphysical worry here too. I have argued that invariant causal influences together with laws of composition can ground applications of the reductive method. But if we allow any old laws of composition, then it seems we will be able to ground the reductive method in all circumstances, including circumstances where we would intuitively think the method was inappropriate. To see why this is so, note that a law of composition is simply a function that takes component influences as inputs and then outputs some means for predicting the resultant behaviour. But given *any* set of behaviours and *any* assumed component influences, there will be a function that maps those components to the appropriate behaviour. Unless there are some restrictions on the form that a composition function can take, the reductive method will be rendered trivial, for we will be able to analyse any behaviour in terms of whatever component influences we wish. I will consider the constraints on laws of composition in Chapter 5.

3.7 Fundamental Influences

The arguments above suggest that the reductive method assumes the existence of something which plays the role of causal influence, and that this role cannot be filled by anything in the standard ontologies. One might wonder, then, whether causal influences are a fundamental, *sui generis*, constituent of reality, or whether they can be constituted by elements of reality that are not themselves influences, and presumably, are not themselves included in standard ontologies (I discuss the separate question of whether influences can be constructed from other, more basic influences in Chapter 6).

If reality were as described by classical physics, then it would be very tempting to suppose that the most basic influences are the fundamental forces, and that these are indeed *sui generis* constituents of reality. However, classical physics is no longer our best description of the fundamental nature of the world; that distinction is held by the Standard Model of quantum field theory (QFT). So the question, then, is whether QFT includes fundamental influences, or whether it shows how influences (in this case forces) can be constituted by elements that are not themselves influences. Unfortunately, I do not think that there is any clear answer to this question at the moment. The metaphysics of QFT is a notoriously difficult topic, and there is no consensus on what it is telling us about the world.

One might be worried, however, that QFT actually contradicts the arguments I have given here. For QFT is commonly described as telling us that the fundamental forces are constituted by the exchange of subatomic particles. The electromagnetic force, for example, is said to be constituted by the exchange of photons (Hobson, 2011, 13; Peskin and Schroeder, 1995, 125). If this is the case, then it would seem that a force can be constituted by the behaviour of particles, and hence that influences can indeed be constituted from elements found in standard ontologies. But this argument from QFT rests on a misconception. Even if it is true that forces are constituted by the exchange of particles, these 'particles' are not like the objects that are found in standard ontologies. For QFT tells us that subatomic particles can also be viewed as 'field excitations', and, unlike ordinary particles, these field excitations can be added together (or *superposed*) in a similar way to forces.

For the moment, then, the fundamental ontology of forces, and hence influences in general, remains unresolved. It is possible that the most basic influences are constituted from more fundamental ingredients that are not themselves influences. If this is the case, however, then these fundamental ingredients are unlikely to be the kind of thing we find in standard ontologies. Note also that the reductive method would not be applicable at this fundamental level. On the other hand, it is possible that influences are a fundamental part of reality, in which case it is likely that the fundamental influences include the fundamental forces—or whatever their equivalent is in QFT. Non-fundamental influences would then be composed from these fundamental influences, as described in Chapter 6.

But note that fundamental influences need not all be forces. According to the four-dimensionalist view of the universe (as advanced by Ted Sider 2001), for example, the continued existence of an individual through time involves a string of three-dimensional parts arranged appropriately through time, with the collection of these parts forming the four-dimensional individual. If the appropriate arrangement of these parts (and hence the continued existence of the individual) is not to be a matter of complete coincidence, then the existence of a three-dimensional object (or field excitation) at one point in time and space must have implications for the existence of a similar object at some nearby point in space and time. A natural way to make sense of this in the ontology I am developing here is that the existence of a three-dimensional object at a given time produces an influence directed toward the existence of a similar three-dimensional object at a nearby time. Such an influence would not be a force, since it is not directed towards an acceleration.[14]

3.8 Conclusion

What I would have liked to have done here is produce a transcendental argument for the existence of causal influences. Such influences, I would like to say, are necessary presuppositions of one of the most useful and successful (and I might suggest essential) methods we have for understanding the world. The difficulty, of course, is proving necessity. Perhaps there are ways to make sense of the reductive method without recourse to causal influences and I simply lack the imagination to think of them. What I have shown is that the method requires some constant element or other, and that the usual suspects (behaviours, laws, dispositions, and powers) cannot play the role required, whilst causal influences can.

Causal influences are no ontological free lunch (to steal a term from Armstrong 1997), so I can see the temptation to do without them. But if I am right that causal influences are assumed by the reductive method, then the spectacular success of this method gives us good reason to believe in them. We shall see through the rest of this book that an ontology that includes causal influences and causal powers can do more than simply ground the reductive method. Influences can help us solve problems in understanding laws of nature, causation, the possibility of emergent properties, and maybe even ethics. Before turning to these issues, however, it will be useful to develop a more detailed account of the metaphysics of power and influence, and of the role they play in the reductive method. That is the task of the next two chapters.

---

[14] Also, in the language of Chapter 5, this influence would be of rank one, whilst forces are of rank two.

4

Causal Power

In the previous chapter, I introduced the concept of causal influence, but I also made extensive reference to causal powers. The aim of this chapter is to investigate the metaphysics of causal powers that was assumed in my discussion of the reductive method, and in particular to investigate the relation between causal powers and causal influence.

It is a common thought that causal powers are dispositions. Authors such as George Molnar (2003), and Stephen Mumford and Rani Anjum (2011a, 4), for example, use the words 'power' and 'disposition' interchangeably. Others, such as Rom Harré (1970), Brian Ellis and Carolyn Lierse (1994), and Alexander Bird (2016), argue that powers are a particular kind of disposition. I too will argue that causal powers are a kind of disposition. In particular, I suggest that powers are dispositions to exert influences.

4.1 Powers as Dispositions

A disposition is a tendency to manifest some characteristic behaviour in some appropriate context. A fragile glass, for example, is disposed to break when struck. What has traditionally caught philosophers' attention is the fact that an object can possess a dispositional property like fragility without that property ever being manifest. If a fragile glass is never struck in the right way, then it may never break; yet its failure to break does not lessen its fragility. Dispositions, it seems, have more to do with the *possible* behaviours of objects than with their actual behaviour.

It is generally agreed, then, that dispositions have an interesting modal character. The traditional way to spell out the modal nature of dispositions (first suggested by Ryle (1949)) involves providing an analysis in terms of a subjunctive conditional. The simplest version of such an analysis is the *Simple Conditional Analysis*:

Definition 5 (SCA). x is disposed to manifest M in response to stimulus S iff were x to undergo S, x would yield manifestation M.

It is now almost universally accepted that this analysis fails, since the biconditional is made false by the possibility of *finks*, *antidotes*, and *mimics*.[1]

---

[1] Though Choi (2006; 2008) and Gundersen (2002) reject this almost universal claim.

An object's disposition is said to be *finkish* if things are set up so that the occurrence of the stimulus that triggers the disposition (*S* in the SCA) also causes the object to lose the disposition in question. David Lewis (1997), for example, has us consider a sorcerer who takes a liking to a particular fragile glass. The sorcerer watches the glass carefully and resolves that if the glass is ever struck, he will cast a spell that changes the structure of the glass in such a way that renders it non-fragile. Before being struck, the glass is just like any other fragile glass that came from the same production line, so surely it too is fragile. But if the sorcerer can act fast enough to remove the disposition if ever the glass is struck, then it will be false that the glass will shatter when struck, and we have a counterexample to a simple conditional analysis of the fragility of the glass.

An *antidote* (also known as a *mask*), on the other hand, is something that leaves the disposition intact, but interferes with the causal chain between the disposition and its manifestation in such a way that the manifestation does not come about. Bird (2007, 28), has us imagine a sorcerer very like Lewis's. Bird's sorcerer, however, resolves to protect the glass if it is ever struck, not by removing the glass's fragility, but rather by interrupting the process that leads from striking to shattering. The sorcerer might do this, for example, by administering shock waves to the glass that precisely cancel out those caused by the original striking. In this case the glass retains its fragile disposition, but still does not break when struck, and so once again provides a counterexample to the SCA.

Finks and antidotes provide counterexamples to the left-to-right direction of the biconditional in SCA. *Mimics*, on the other hand, provide counterexamples to the right-to-left direction. They are cases in which an object does not have the disposition in question; yet the manifestation *M* nonetheless occurs upon stimulus *S*. Styrofoam bowls, for example, are not fragile in the same way that a glass is: they do not break into many pieces when struck. But imagine that a styrofoam bowl is struck in the presence of the Hater of Styrofoam. Upon hearing the telltale sound of styrofoam being struck, the Hater of Styrofoam is alerted to the presence of the bowl and immediately tears it to pieces. In the presence of the Hater of Styrofoam, a styrofoam bowl will break into pieces when struck, but nonetheless, it does not posses the disposition of fragility that is possessed by a fragile glass (this example is due to Lewis 1997).

Despite general agreement that the SCA fails, it is still a common starting point for discussions of dispositions, and many philosophers have attempted to provide conditional analyses of dispositional properties by introducing complications to the SCA that are designed to avoid counterexamples involving finks, antidotes, and mimics—see, for example, Prior (1985); Lewis (1997); Fara (2005); or Manley and Wasserman (2008). This trend in the literature suggests that conditional analyses like the SCA give us, at least, a rough and ready handle on the modality that philosophers believe is characteristic of dispositionality.

4.1.1 Dispositional Modality

Recently, however, Mumford and Anjum (2011a) (taking up an idea from Molnar 2003) have argued that dispositions are characterized by a very different kind of modality. Imagine that you arrive home one winter's evening to find the house cold and dark. Fortunately, you had the foresight to leave a fire set in the fireplace, so you light the fire, sit back with a glass of red wine, and relax as the fire warms your skin and the room around you. It seems clear in this case that the fire has caused the room to get warmer.

Mumford and Anjum argue that there seems to be a simple and natural way to understand causation like this in terms of the activity of the powers of the objects involved. The fire has the power to warm the room, which is to say that it is disposed to warm the room when it is burning. The fire's causing of the room to become warmer, then, is just a matter of the fire manifesting its disposition to warm the room. In general, the simple idea behind a dispositional account of causation is that all effects are the result of the powers that are at play in the situation—causation just is the manifesting of dispositions.

Let us suppose that the room you are in has a large picture window overlooking your garden. The window provides a nice view, but on a night like this it also tends to cool the room by letting heat escape. In this situation there are two powers at play: the power of the fire to warm the room, and the power of the window to cool the room. In such a situation it is natural to think that what actually happens to the temperature of the room (whether it increases, decreases, or stays the same) will be a result of both these powers acting together. Fortunately, the window is double-glazed, so that its power to cool the room is weaker than the fire's power to heat it, and, thankfully, the room will get warmer.

But now imagine that your 7-year-old daughter leaves an external door open. The open door adds a third power to the mix, and the change in room temperature will be a result of the action of all three powers. If it is cold enough outside, then the combined powers of the door and the window to cool the room may outweigh the power of the fire to warm it, and so the room will cool down.

The conclusion that Mumford and Anjum draw after considering situations like that above is that the ultimate behaviour of a system at any instant will be the net result of all the powers that are acting on the system at that time. Powers, on this view, make only a *contribution* to an outcome, and these contributions must be combined to produce effects. So far, this is all consistent with the concept of powers that was utilized in the discussion of the reductive method of explanation in Chapter 3.

Examples like that involving the fire and the window above also suggest that in order to make sense of the composition of such contributions we need to make use of the concepts of the *strength* and *direction* of the contribution. The fire and window, for example, make contributions in opposite directions, while the window and the door make contributions in the same direction. It is also important that the window's contribution is weaker than the fire's, but the combined contribution of the window and the open door is stronger than the contribution of the fire.[2]

Mumford and Anjum (2011a; 2011b) use this insight to motivate a graphical 'vector model' of causation in which the contributions of active powers are represented by arrows whose length represents the strength of the power, and whose direction represents the outcome towards which the power is disposed. For example, we could use an arrow pointing to the left to represent a disposition to warm the room, and an arrow pointing to the right to represent a disposition to cool the room. The situation described above could then be represented as in Figure 4.1. If we assume that the contributions represented by the arrows here add according to the familiar parallelogram rule of vector calculus—according to which you place the tail of the second arrow on the head of the first and the resultant is then represented by an arrow from the tail of the first vector to the head of the second—then diagrams like this one can be very useful in helping make sense of what is going on. In the situation depicted, for example, the combined effect of the window and the open door is stronger than the tendency of the fire to warm the room, and so the result is a tendency for the room to cool.

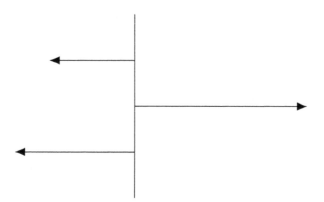

Figure 4.1. A simple vector model of causation

---

[2] But remember that, as discussed in Chapter 3, the concepts of strength and direction are not as straightforward as one might think.

This graphical representation is very nice, but it does have limitations. Most importantly, the vector model is less useful if we cannot assume the parallelogram rule of vector addition. Without this rule we cannot use the graphical representation to read off resultants by stringing arrows together head to tail. Mumford and Anjum themselves point out that there are many cases in which the composition of powers is not linear and hence does not obey the parallelogram rule. As an example, they note that while eating one chocolate bar might incline one towards happiness, eating ten chocolate bars in quick succession would have the opposite effect (2011a, 86).

Bird (2016) argues that the possible non-linearity of causal composition is not the only problem for the vector model of causation. He argues, for example, that contrary to Mumford and Anjum's claim, the vector model cannot be extended to multiple dimensions. For example, if the fireplace were disposed to both warm and dry the room, while the open door was disposed to cool and humidify the room, then we might think that we could represent the contributions of the fire and the window in a two-dimensional space, with one axis representing changes in temperature and the other representing changes in humidity. But, Bird argues, there is no way to think of this two-dimensional space as a vector space, since there is no sense to the idea of 'rotating' a vector that originally points along the temperature axis so that it then points along the humidity axis. Less metaphorically, the problem is that there is no sensible metric that encompasses both temperature and humidity. How can we compare the strength of a tendency to heat the room with the strength of a tendency to humidify it? Bird concludes that the vector model of causation is only applicable in the simplest one-dimensional cases, and, he says, in those cases it does not add anything to existing models.

I am not nearly so pessimistic as Bird regarding the vector model of causation. Despite the limitations of the model, Mumford and Anjum put it to good use, especially when making sense of the possibility of causation by absence (2011a, 143–7). Nonetheless, I will suggest a very different way of modelling causal contributions in Chapter 9, and I believe that this new model will avoid Bird's criticisms of the vector model.

My aim at the moment, however, is not really to discuss Mumford and Anjum's vector model of causation, but rather, to consider the notion of dispositionality that underlies the model. As Mumford and Anjum repeatedly stress, powers, even when active, do not necessitate their effects (logically, metaphysically, or nomologically). It is rarely, if ever, the case that if power A is active, effect E must follow: whether E follows will depend on what *other* powers are at play. So, for example, even though the fire is disposed to heat the room, it is not the case that lighting the fire will necessarily result in the room getting warmer. If it is cold enough outside and I leave the door open, then the fire may do no more than slow the decrease in temperature. According to Mumford and Anjum, then, causation is not necessitation. This sets the view apart from the kind of view championed by

Armstrong (1997) according to which causation is a relation in which the instantiation of some universal F necessitates the instantiation of some universal G.

On the other hand, a power does not treat all possible outcomes equally: it points towards one in particular. When we say that the fire is disposed to warm the room, we do not merely mean that it is possible the room will get warmer if we light the fire. After all, it is also possible the room will get cooler if we light the fire (if we leave the door open, for example). The directionality of the fire's power—the fact that it 'points towards' warming rather than cooling—seems to involve a modality that lies between necessity and mere possibility. Mumford and Anjum call this new modality 'dispositional modality' and argue that it is a primitive notion that cannot be analysed in terms of mere possibility and necessity (see Anjum and Mumford, 2011; Mumford and Anjum, 2010; 2011a,b).[3]

Notice that Mumford and Anjum's dispositional modality is not the modality that the SCA tries to capture. True, in both cases the modality involves some kind of 'directedness' towards a particular outcome. But the conditional analysis just says that this outcome will occur if certain conditions are satisfied and says nothing about what happens otherwise. Mumford and Anjum's dispositional modality, on the other hand, involves a directedness that contributes to what goes on even in cases where the outcome to which it is directed does not occur.

Mumford and Anjum are right, then, that dispositional modality is something new. But this also means that it is not the modality that is associated with the usual notion of a disposition (and so one may doubt that it deserves the name 'dispositional modality'). One may wonder, then, how these two modalities relate to each other. How does Mumford and Anjum's notion of dispositional modality connect to the traditional notion of a disposition? This is not a question that Mumford and Anjum address explicitly, and perhaps their thought is that the concept of dispositional modality is intended to *replace* the traditional notion of dispositionality. I would like to suggest a different answer. If, as I suggested above, a power is not just any old disposition, but, rather, a disposition to exert an influence, then we can think of the traditional modality as governing the *input* to a power, whilst Mumford and and Anjum's dispositional modality has to do with a power's *output*. On the input side, a power is a property that will only be active if the right conditions occur. On the output side is the fact that, even when active, a power does not necessitate its effect, but merely contributes to it.

If a power is a disposition to exert an influence, then the manifestation of a power is not typically a change in some property, but, rather, the existence of a causal influence that makes a contribution to such a change. We thus get the threefold ontology mentioned in Chapter 3, consisting of (1) powers, (2)

---

[3] Actually, I believe that there is more necessitation in dispositional modality than Mumford and Anjum accept. I will discuss this further in Chapter 8.

influences, and (3) states. The general picture is then as follows: the state of a system triggers various powers, these active powers exert influences, and the influences then combine to produce a change in the state of the system. As mentioned in Chapter 3, a picture like this has been endorsed by Molnar (2003, 195), Cartwright (2009, 151) and Marmadoro (2017). Jennifer McKitrick (2010) also attributes this kind of picture to Mumford and Anjum. However, it is not clear that such an attribution is correct, since, as I mentioned above, it seems likely that Mumford and Anjum intend the concept of dispositional modality to replace, rather than augment, traditional analyses of dispositions.

I speak here of a disposition being 'triggered' and one might object, as does E. J. Lowe (2011), that some dispositions require no stimulus at all. Consider, for example, the disposition of a radium atom to undergo spontaneous radioactive decay. The manifestation of this disposition, says Lowe, is completely insensitive to all external conditions. Now, Marmadoro (2016, 205–6) tells us that Lowe is simply wrong about this, stating that the so-called 'spontaneous' decay has been shown to be triggered by 'vacuum fluctuations'. But the situation is more complex than Lowe or Marmadoro suggest. First, the research that Marmadoro points to (Yokoyama and Ujihara, 1995) is concerned with the 'spontaneous' emission of photons by excited atoms, rather than radioactive decay of nuclei. Second, these models of spontaneous emission involve both a contribution from vacuum fluctuations and a contribution due to an electron's interaction with its own field (Dalibard et al., 1982). But even if Lowe is right about the existence of dispositions that have no triggering conditions, such dispositions can easily be accommodated within the framework I have developed here—we simply have to allow relatively trivial stimulus conditions. In particular, the triggering condition for such a power will be the mere possession of that power (in the language to be developed in Chapter 5, this is a 'rank one' power). So, for example, we could explain truly spontaneous decay of a radium atom by positing that the radium atom has a power to exert an influence on itself that is directed toward decay, and the stimulus condition for this power is simply being a radium atom. This condition is satisfied by all radium atoms at all times. The reason that the decay may or may not happen at any given time is that either the power or the influence it manifests is probabilistic in nature (see below for a discussion of such probabilistic powers and influences).

It should be clear that the analysis of causal power I offer here is not reductive, since I am explaining a causal concept 'power' in terms of two other causal concepts: 'exerting' and 'influence'. Like Mumford and Anjum, I take the output of an active power (an influence) to involve a primitive modality that cannot be analysed away, and so a reductive analysis is not possible. I justify the use of causal terms to denote this primitive element in Chapter 8.

4.1.2 McKitrick's Objection

McKitrick (2010) has argued that threefold ontologies like the one I advocate here are seriously problematic. If the manifestation of a power is not an 'effect' (that is, a change in, or maintenance of, a system's state), but a 'contribution' to an effect, McKitrick asks, what kind of thing is this 'contribution'? She rightly notes that the contribution cannot be an intermediary event that occurs between the power and the effect, since this would simply be an intermediary effect and the view would not be an alternative to the view that manifestations are effects. She then goes on to consider three possible answers to the question: the contribution is part of the effect, the contribution is a property of the effect, or the contribution is a force. I will not dwell on the first two of these possibilities, since neither corresponds to the position that I am putting forward here. Suffice it to say that McKitrick argues convincingly that neither of these possibilities makes any sense. It is the third possibility that is relevant to the picture I have drawn. My position is that the manifestation of a power is an influence and I have argued that forces are paradigmatic examples of influences. So let us turn to McKitrick's argument against the ideas that the manifestation of the power is a force.

Consider a case in which an apple falls to the Earth. According to the view that I am advocating, the motion of the apple is not a manifestation of the Earth's power of gravitation. Rather, the manifestation of the Earth's gravitational power is a force which then contributes to the motion of the apple. McKitrick complains that this misdescribes the situation. The right thing to say, she insists 'is not that gravitational power produces gravitational force, but that gravitational power just is gravitational force' (2010, 82). The problem, according to McKitrick, is that forces seem very much like powers, and so it seems artificial to distinguish between gravitational power and gravitational force. If we, nonetheless, insist on distinguishing powers from the forces that they manifest, she says, we just get ourselves into more trouble. Insisting on a distinction between powers and forces does not make forces any less power-like, and so McKitrick concludes that distinguishing a power from the force it manifests is to add a second power to the picture. When the first power is triggered, it manifests a second power which then contributes to the effect.

Not only is this picture needlessly complex, but a regress is looming. If all powers only manifest contributions to an effect, and if all contributions are forces, and if forces are themselves powers, then it it is true of all powers that they can only contribute to an effect by manifesting a second power which then contributes to the effect. But, by this same reasoning, the second power can only contribute to an effect by manifesting a third power which can then only contribute to an effect by manifesting a fourth power, and so on. McKitrick concludes that we are faced with a situation reminiscent of Zeno's paradox: 'Infinitely many manifestations/forces must exist before the triggering of a power can result in an effect' (2010, 83).

As with Zeno's paradox, it is not clear that such an infinite regress is impossible, and the supporter of manifestations-as-contributions could simply dig in and defend the admittedly uncomfortable position that the manifesting of a power always involves an infinite regress. Fortunately, however, we need not take this route, for McKitrick's argument is flawed.

McKitrick's argument relies on the premise that forces are themselves powers. The only reason she gives for believing this premise is that 'forces seem very much like powers' (2010, 82), and she does not provide any details regarding the way in which forces are similar to powers. Presumably she takes the similarity to be intuitively obvious; perhaps because 'force' and 'power' are typically taken to be dispositional/causal concepts. But my intuitions, at least, suggest that force and power are completely different. Gravitational power, for example, is captured in physics by our concept of gravitational mass, which is represented by a scalar quantity, while forces are represented by vectors. Intuitions aside, the picture I have developed here draws a sharp distinction between powers and influences. Powers are essentially dispositional and thus involve a modality that is characterized by something like the SCA. Influences, on the other hand, are characterized by the distinct, though similarly named 'dispositional modality'. Powers, we might say, involve an 'input modality', while influences involve an 'output modality'. Perhaps the term 'dispositional modality' obscures this difference between these modalities. According to the picture developed here, dispositional modality characterizes the *manifestation* of a disposition rather than the disposition itself. Thus, it might be better named 'contribution modality' or 'influence modality'. Because powers and influences are characterized by different modalities, they must be different things. Influences (and forces in particular), therefore, are not powers and there is no regress. McKitrick's argument against causal influence fails.

4.1.3 No Antidotes

If, as I have argued, a power is a disposition to manifest an influence, then we might attempt a simple conditional analysis of powers as follows:

Definition 6 (SCAP). x has the power to exert influence I in response to stimulus S iff were x to undergo S, x would manifest influence I.

So, for example, the Earth has the power to attract nearby apples since, if there is an apple near the Earth, the Earth will exert an attractive force on the apple.

SCAP is modelled directly on the simple conditional analysis of dispositions, and so we might expect that it will fall foul of the same counterexamples that plague the SCA. Fortunately, however, this is not the case. Suppose, for example, that we were to give a dispositional gloss of gravitational mass, of the same kind that Bird (2007, 45) gives for electric charge. Then, we would say that gravitational mass just is the disposition to attract other masses. Now, there is an ambiguity in the word 'attract' as used here. On the one hand, we could say that x attracts y just in case x produces an acceleration of y towards x. If we read 'attract' this way, then the manifestation of the Earth's gravitational power will be an acceleration of massive objects towards the Earth's centre of gravity. The problem with this account of gravitational power is that most massive objects do not accelerate towards the Earth, not even those that are in close proximity. Tables, chairs, buildings, trees, and rocks, for example, typically rest motionless on the surface of the Earth rather than accelerating towards it. Thus, there are numerous counterexamples to a simple conditional analysis of the Earth's gravitational power as a disposition to produce accelerations in objects that are placed in proximity. The disposition to produce such accelerations is almost always interfered with, and so a simple conditional analysis of this sort will almost always be faced with antidotes.[4]

On the other hand, we could take the word 'attract' to signify the presence of a force. So x attracts y just in case x produces a force on y that is directed towards x.[5] In this case, contra McKitrick, the manifestation of the Earth's gravitational power would be a force rather than an acceleration. The claim that gravitational power is a disposition to exert a force does not suffer from the same counterexamples as the claim that it is the disposition to produce an acceleration. Although tables, chairs, buildings, trees, and rocks typically do not accelerate towards the Earth, they do feel a force directed towards the Earth.[6] Indeed, Newtonian physics tells us that every mass in the universe will be subject to a force due to the Earth's gravitational power. Thus, according to Newtonian physics there will be no counterexamples to a simple conditional analysis of the Earth's gravitational power to manifest forces.[7]

The simple conditional analysis of powers can avoid many of the counterexamples to the simple conditional analysis of dispositions more generally, because the manifestation of a power is an influence, and influences *combine* to produce an outcome. The influence manifested by a power may be invariant across a large range of contexts even though the behaviours to which the influence contributes may vary across these contexts. In general, a simple conditional analysis of some power will avoid counterexamples to the extent that the influence manifested by the power in question is invariant. In fundamental physics, we posit powers that are invariant across *all* contexts: given the stimulus condition, the power will exert a particular influence no matter what else is going on. In fundamental physics, then, we posit powers that cannot be finked, and have no antidotes. Bird (2004)

---

[4] It was this problem that led Nancy Cartwright to famously claim that the laws of physics lie. See Chapter 7 for further discussion of the laws of nature.

[5] More correctly, x produces a force on y that is directed towards an acceleration of y towards x.

[6] I have here assumed that component forces exist. For a justification of this assumption, see Chapter 6.

[7] According to relativity, of course, gravitational influences are not instantaneous, and so only events that are timelike separated will influence each other. But if we build this limit into the stimulation conditions for gravitational powers, then there will be no counterexamples.

has argued that there can be no finkish dispositions at the fundamental level, and that antidotes at the fundamental level are at least likely to be less widespread than at the macro level. But if we understand powers as I have suggested, then physics tells us that there are *no* antidotes to powers at the fundamental level.[8]

But what of powers outside fundamental physics? I have suggested that the reductive method of explanation can be applied in many different contexts. Not only do we use the method to explain the behaviour of collections of physical particles, but we use it in explanations of population cycles in terms of interactions between predator species and prey species, explanations of financial crisis in terms of the influence of sub-prime loans, and so on. If any of these explanations are interesting in the sense I described in Chapter 3 (and, I believe they are), then these explanations must assume powers that manifest invariant influences, powers that are not susceptible to finks and antidotes. But note that these powers need only be free of finks and antidotes within the range of circumstances to which the explanation is applied. When explaining population cycles in terms of interactions between predator species and prey species, we assume that, given certain facts about the size and location of the two populations, the influence each has on the other is invariant. This assumption may be correct in a wide range of natural circumstances, and a reductive explanation based on this assumption could then be correct in these circumstances. However, there may be circumstances in which this assumption of invariance fails, such as cases where the predator species are domesticated (and so feel less desire to hunt).

Similarly, a successful reductive explanation only requires that the influences manifested by the powers to which the explanation refers are *approximately* invariant. If the influence that a predator population exerts on a prey population depends in some very small way on some background condition that has not been taken into account, this lack of invariance may be irrelevant to the success of a reductive explanation of population cycles, so long as the deviation from invariance makes no practical difference to the phenomenon being explained.

Putting these thoughts together, we can modify SCAP to allow for approximate powers as follows:

Definition 7 (SCAP*). In context C, x has the power to exert influence I in response to stimulus S iff were x to undergo S in context C, x would manifest an influence sufficiently close to I.

Here the context C specifies certain background conditions that are assumed to hold as well as specifying what counts as 'sufficiently close to I'. A *true power*, such as we find in fundamental physics, is one in which SCAP* holds in all possible

---

[8] See Section 7.2 for a more detailed discussion of the possibility of fundamental antidotes.

contexts. A reductive explanation, however, only requires powers that approximate a true power in the contexts that are relevant for the explanation.

Note that if we find ourselves attempting to apply the reductive method in a context in which the powers are not even approximately invariant, we need not, and typically do not, abandon the reductive method. Rather, we can take the failure of the method to indicate that the powers that were referred to are not basic. We then search for a set of more basic components whose powers are invariant within the range of circumstances that we wish to explain. Note that this procedure is not peculiar to the special sciences: the same thing happens in physics. In normal circumstances, for example, we can treat the Earth as a basic object that has a basic power to exert a 9.8 N force on any 1 kg mass that is near the Earth's surface. This power remains invariant over a large range of circumstances, from situations in which the Earth and 1 kg mass exist in isolation to situations like the actual one in which the Earth is part of a more complex solar system. Indeed, the power will remain invariant even in some very extreme variations of the actual situation. The Earth would have the same influence on a nearby mass even if the Earth were in orbit around a binary star or were one of the moons of Jupiter. However, if the two-object system consisting of the Earth and 1 kg mass were placed close enough to a black hole, then tidal effects would stretch the Earth out of its roughly spherical shape, and possibly tear it apart. In such a situation we could no longer treat the Earth as a basic object with the basic power described. But this does not mean we cannot apply the reductive method: we must simply take smaller pieces of the Earth as our basic objects. This is precisely what happens in computer simulations of interactions between a planet and a black hole, for example.

In general, when a reductive explanation for some phenomenon fails, it is often useful to look for a more fine-grained explanation. Interestingly, as we go more fine-grained, we tend to find powers that more closely approximate true powers. The end result of this process of fine-graining is fundamental physics, with true powers that cannot be finked and have no antidotes.

4.1.4 Dealing with Mimics

Can SCAP or SCAP* be defended against counterexamples involving mimics? I think that they can, but the defence does not depend on the nature of influences, and so it is equally available to SCA.

Consider again the case of the Hater of Styrofoam. Why do we intuitively say that the styrofoam bowl is not disposed to break when struck? After all, if you strike the bowl, it will indeed break. Lewis's response is that the manifestation of fragility is not merely breaking, but breaking through a certain *direct and standard process* (Lewis, 1997, 153). The styrofoam plate might break if struck, but the process by which the breaking happens involves the Hater of Styrofoam, and so is neither direct nor standard. In a similar vein, Gabriele Contessa (2014) suggests that the difference between a mimic and a disposition is that, in the case of a mimic, the manifestation in question occurs due to an 'external influence'—the Hater of Styrofoam in this case.

The general thought, I suggest, is that we refuse to attribute a disposition to a system in cases where the causal process leading from stimulation to manifestation involves elements that are both external to the system in question and not part of the standard background conditions. But since the conditions that count as standard can vary from one context to another, whether or not the truth of the right-to-left conditional in SCA is due to a dispositon or a mimic will also vary from one context to another. If the hater of syrofoam were a part of the standard backgournd conditions, then we would indeed say that the syrofoam plate is fragile.

If I am right about the role of a standard context in determining whether something is a disposition or a mere mimic, then we can distinguish the two cases by defining a *standard disposition*, which invokes the same notion of 'context' that is referred to in SCAP*, but restricts attention to standard contexts:[9]

Definition 8 (SCA**). x has the *standard disposition* to manifest M in response to stimulus S iff were x to undergo S in any standard context, x would yield manifestation M.

Although the styrofoam plate is disposed to break when struck in the non-standard context that involves a hater of stryofoam, it is not disposed to break in all *standard* contexts, and so we do not attribute to the plate the standard disposition.

We can make a similar change to SCAP* in order to avoid the problem of mimics in the case of powers:

Definition 9 (SCAP**). x has the *standard power* to exert influence I in response to stimulus S iff were x to undergo S in any standard context, x would manifest an influence sufficiently close to I.

Note that a true power—one for which SCAP* holds for all contexts—will necessarily be a standard power.

4.2 Multi-Track Powers

The analysis of powers embodied in SCAP assumes that these powers are *single-track* dispositions—they have a single stimulus condition and a single manifestation. Some powers, however, seem more complex than this. The influence that one negative charge exerts on another, for example, depends on the distance between

---

[9] This strategy is very similar to that employed by Choi (2005).

them. Negative charge, therefore, seems to be a *multi-track* power, having more than one kind of stimulus or manifestation.[10]

To completely characterize a multi-track disposition we will need more than a single conditional. One possibility is to describe a multi-track disposition with a conjunction of conditionals. Bird (2007, 23), for instance, has suggested that all fundamental properties will be single-track dispositions, while multi-track dispositions are complex properties composed from the simple fundamental dispositions. In this case it seems entirely appropriate to describe a multi-track disposition using a conjunction of conditionals.

But let us take a moment to examine Bird's suggestion. Electric charge seems like a good candidate for a fundamental property; yet it seems best understood as a multi-track power.[11] Consider, for example, the property of unit negative charge, which I will label Q^-. We can describe Q^- as the disposition to exert a force F on another charge q at distance r, where, for any values of q and r, the magnitude of the force is given by:

$$F = \frac{\kappa_e q}{r^2}$$

where κ_e is Coulomb's constant, and a positive value of F indicates a force directed towards the original particle.

Because q and r are real-valued variables, the disposition associated with unit negative charge can give rise to an infinite number of different forces, depending on the precise stimulus conditions. Thus, unit negative charge seems to fit the bill of a multi-track disposition.

Following Bird's suggestion, however, we could analyse unit negative charge as an infinite conjunction of single-track dispositions each describing the force that will be exerted in a particular situation. So, for example, given any values of x, y, and z, we could define the single-track disposition $Q_{x, y, z}$ to be the disposition to exert a force of z newtons on a charge of x coulombs at a distance of y metres. Then Q^- is just the conjunction of all and only the dispositions $Q_{x, y, z}$ such that

$$z = \frac{\kappa_e x}{y^2}$$

The problem with this suggestion is that if the fundamental powers are single-track dispositions like $Q_{x, y, z}$, then it is somewhat mysterious that all charged particles have ended up with an infinite bundle of dispositions that just happen to conform to Coulomb's law. Why this set of dispositions rather than a set conforming to an inverse cube law? Indeed, why not a purely random set of dispositions,

---

[10] Gilbert Ryle (1949) was the first to explicitly discuss multi-track dispositions. He suggests, for example, that knowing French is dispositional, and has many different stimuli and manifestations (reading French, writing French, speaking French, etc.).

[11] One might object that electric charge has manifestations that differ in degree not in kind, and hence is not multi-track. Following Bird, I will understand 'multi-track' to include such dispositions.

or—perhaps more worrying—an incomplete set, that includes no disposition to exert a force in the presence of a charge of, say, 203 coulombs? If the fundamental dispositions are single-track, then one bundle of such dispositions seems no more likely than any other.

If, on the other hand, we posit that unit negative charge is a fundamental multi-track disposition, then we are not faced with these questions. Charged particles conform to Coulomb's law because to possess a charge *just is* to possess a disposition to exert a force in accordance with Coulomb's law.

One may respond that explanation has to bottom out somewhere, and it is simply a brute fact that, in our universe, particles tend to come with bundles of single-track dispositions that conform to Coulomb's law. Indeed, one might point out that even if the multi-track Q^- is the fundamental disposition, the only explanation for Coulomb's law is the brute fact that in our universe some particles happen to possess dispositional property Q^-, rather than, say, Q^*, which conforms to an inverse cube law. However, the two explanations are not equally unsatisfactory. If Q^- is fundamental, then Coulomb's law is explained by one brute fact—the instantiation of Q^-. But if it is the single-track dispositions $Q_{x,y,z}$ that are fundamental, then Coulomb's law is explained by reference to an infinite number of brute facts.

Perhaps more worrying for Bird is the following difference. In his book *Nature's Metaphysics*, Bird argues that the fundamental laws of nature are a consequence of the fundamental dispositional properties that objects possess. We can see how this would work if the multi-track power Q^- were fundamental, for then Coulomb's law would be true of all particles that possessed Q^- simply because that is what it means to possess this property.[12] However, if $Q_{x,y,z}$ are the fundamental properties, then the regularity that needs explaining is that the dispositions $Q_{x,y,z}$ are related to one another in a way that conforms to Coulomb's law. But there does not seem to be any fundamental disposition that can explain this. Thus, restricting fundamental dispositions to single-track ones like $Q_{x,y,z}$ means that fundamental laws like Coulomb's cannot be explained in terms of fundamental dispositions, contrary to Bird's main thesis.

In general, I would suggest, any attempt to explain a law-like regularity by reference to a bundle of dispositions just shifts the question from one domain to another. For, in order to produce the appropriate regularity in their manifestations, the dispositions in the bundle will need to satisfy a directly analogous regularity among themselves. Hence, all else being equal, it will always be preferable to posit a single multi-track power to explain a law rather than a bundle of single-track powers.

One might object that if positing a single multi-track power is always preferable to a bundle of powers, then we should be treating all of the powers of an object as

---

[12] The relation between laws and powers is discussed further in Chapter 7.

different aspects of a single, rather complex, multi-track power. However, I only claimed that a single multi-track power is preferable if all else is equal. There is one (and perhaps only one) consideration that can push against positing a single multi-track power. It is simpler, and more explanatorily useful, to posit a bundle of powers rather than a single multi-track power when different objects possess different subsets of the bundle. This is why, for example, we separate the power of gravitational mass from the power of electric charge rather than treating them as a single, complex, multi-track power possessed by all massive charged objects: there are objects that have the powers associated with mass, but not those associated with charge. Even in this case, however, the separated powers are themselves multi-track, not single-track. Our best explanations, I suggest, will posit the smallest number of multi-track powers that is consistent with separating out powers that do not always appear together.

Finally, one might worry that a power which manifests different influences in different situations will necessarily fail to have the kind of invariance that is required by the reductive method. In fact, this is not so. A power will have the appropriate invariance so long as it gives rise to the same influences in the same situations. The important point for reductive explanation is that 'the same situation' must be taken to involve only a restricted subsystem. So, for example, the electric charge of a particle will manifest different influences on another particle if we change the charge of the second particle or the distance between them. But if we keep the charge and distance between the two particles fixed, then the influence manifested will be the same *no matter what other particles are around* (or, indeed, no matter what else is happening at all). In this way, even a multi-track power can underlie the systems-in-isolation assumption of the reductive method. The relevant notion of invariance is made precise in Chapter 5.

4.3 Functions

Conditionals, such as those that appear in SCAP, are not ideally suited for representing the complexity of a multi-track power, for conditionals associate only a single stimulus condition with a single manifestation. A much better way to represent a multi-track power, I suggest, is with a mathematical function. A function is simply an association between elements in one set (the domain) and elements in another set (the range), such that an element from the domain gets associated with no more than one element from the range. To model powers with functions, then, we let the domain of our function be the set of possible stimuli, while the range of the function is the set of influences that may result. The function then tells us which influence will result from any given stimulus in the domain. To put this another way, powers can be represented as influence-valued functions on the domain of possible stimuli.

Do we really gain anything by thinking in terms of functions instead of con-ditionals? After all, any function is equivalent to a (possibly infinite) string of conditionals of the form 'if element x is chosen from the domain, it will be associated with element y from the range'. In terms of dispositions, this means that any functional description is equivalent to a description in terms of a (possibly infinite) string of conditionals of the form 'if stimulus x were to occur, y would be manifest'. I believe that there are at least four reasons to prefer the functional description of multi-track powers.

First, as Hitchcock (2006, 69) points out, 'the way we choose to represent some phenomenon can shape the way in which we think about that phenomenon'. A description of a multi-track disposition in terms of a string of single-track conditionals surely promotes the kind of single-track view of fundamental dispositions argued against above. I am suggesting that some powers, in particular some fundamental powers, are multi-track dispositions: a single property that can give rise to multiple influences. A representation in terms of multiple conditionals undermines this view, suggesting that there are numerous properties involved in each case, not one. A representation of a multi-track disposition in terms of a single function, on the other hand, reinforces the idea that the disposition is a single property.

A second reason for preferring the functional representation is that math-ematics provides us with numerous tools for investigating and describing the properties of functions. We might, for example, ask if the function representing a power is differentiable, or argue that all fundamental powers must be continuous (manifesting similar influences in similar situations), or consider the implications of a power that is 1:1. If there are relations of interest among possible stimuli and among possible manifestations or possible outcomes, we might investigate the possibility of homomorphisms or isomorphisms between these spaces, making use of various theorems regarding such homomorphisms and isomorphisms (these properties might be of interest if, for example, one was trying to construct a metric on the space of influences based on relations between the stimuli. See Chapter 5 for discussion of such an isomorphism). Of course, all of this can be done starting from a representation in terms of multiple conditionals, but it would be less natural, more complicated, and in many cases would necessarily involve thinking of the set of conditionals as a unified object like a function.

A third reason for preferring the functional representation is that it brings our descriptions of causal powers in line with the kind of descriptions that one finds in quantitative sciences like physics, and hence might help pave the way to understanding the relation between such sciences and the metaphysics of causal powers. The electromagnetic field, for example, is represented in physics by a function that associates each point in space (or space-time), with a vector describing the force that would be felt by a point particle of unit charge at that position. We have already seen that forces can be regarded as a type of causal influence, so the electromagnetic field is an influence-valued function on the set of space(-time) points. So perhaps powers are just a generalized kind of force field ('generalized' because, although forces are causal influences, there may be influences that are not forces). I consider the relation between influence-valued functions and force fields in Section 4.5 below, and show how force fields can be understood as descriptions of multi-track powers.

Finally, the use of influence-valued functions to represent powers naturally captures the invariance that I argued in Chapter 3 is essential for reductive explanation. We will see how this works in the following section.

4.4 Functions and Invariance

If a power is multi-track, then rather than thinking of the stimulus conditions as determining *whether* the power is manifest, we should regard them as determining *how* the power is manifest; different stimuli can lead a power to exert different influences. Now, it may be that there are some conditions under which the power manifests no influence at all, but we can regard exerting no influence at all as just another way for the power to manifest. If we treat powers this way, then the set of all possible stimulus conditions for the power must be the set of all relevant conditions that the power could find itself in, for the power has to 'know what to do' in all possible situations. In other words, the domain of the function representing the power must include the entire state-space (the abstract space of all possible states) of the 'system' that is relevant to the power.

But what is the system that is relevant to a given power? To specify a state-space we need to know what variables can change and which are fixed. Consider, for example, the state-space representing two otherwise isolated electrons. To construct this space we fix the charge, mass, spin, and number of the two electrons, and then consider all the possible values of variables such as the position and momentum of the electrons. However, the same pair of electrons could also be regarded as an example of a more general two-particle system, whose state-space is constructed by fixing only the number of particles and then including all the possible values of mass, charge, and spin as well as position and momentum.

What state-space would we need in order to adequately characterize a power? We might reason that a power has to be 'ready' to act—or not—no matter what the situation, and so go for the largest possible state-space. So, for example, to model unit negative charge we might use a function on the state-space that keeps the negative charge of some particle constant and then varies all other properties, including the position, momentum, number, and type of other particles.[13]

---

[13] We might even need to include 'alien' properties that are not instantiated in this world. See Dumsday (2013) for an argument that alien properties are part of the essence of a disposition.

Taking the largest possible state-space is incredibly ungainly, but there is a bigger problem. If the function representing a disposition has a domain that includes every possible state of a large slice of the universe, then there would be no sense in which the function could be said to represent a property of a local part of this slice like, say, an electron. The function would associate each arrangement of the universe slice with a particular influence, and so would best be seen as describing a disposition of the whole slice. The problem here is essentially the one pointed out by Bertrand Russell (1913) in his famous criticism of causation. If it is only entire slices of the universe that influence later states, then there is no room for a notion of causation as a relation between local states.

Fortunately, many powers can be modelled using a much smaller state-space than that which describes every possible state of an entire universe slice. It is possible to model negative charge, for example, using only a two-particle space. The state-space will include the charged particle whose charge we are modelling, and the particle that is being influenced (or not) by the charged particle. The only variables that will be included in the state-space are the charge and relative positions of the two particles. The power is then represented by a function from this space to the influence on the second particle.

One may wonder how we can get away with such a small state-space. If our model of charge is a function on a two-particle system only, what can it tell us about more complicated situations? If we throw a third particle into the picture, does the model remain silent? The answer is that the model need not remain silent. Any multi-particle system can be considered as a collection of two-particle subsystems—where each possible pairing represents one such subsystem. So, for example, the three-particle system consisting of particles a, b, and c, contains three two-particle subsystems consisting of the pairs (a, b), (a, c) and (b, c). The function representing negative charge will assign an influence to each particle in one of these subsystems based on their charge and relative position—all other properties of the subsystem being irrelevant in considerations of influence due to negative charge. By simply ignoring facts external to the two-particle subsystems, our functional model is applicable in all possible situations. It characterizes a disposition that 'knows what to do' no matter what the world throws at it.

Note that the only reason we can restrict our attention to subsystems when characterizing a power is that we are taking the manifestation of the power to be an influence—something that does not necessitate any given outcome, but rather acts in combination with other influences to produce an outcome. The acceleration of an electric charge, for example, will be influenced by every other charged particle in its backwards light cone. So if the manifestation of electric charge were an acceleration rather than a force, then a complete description of the disposition would have to take into account every possible arrangement of any number of particles in the backwards light cone.

Note also that the restriction of the domain to the state space of a subsystem embodies the assumption that the influence manifested by a power is invariant.

The influence that a power manifests is determined by the state of the subsystem regardless of what is going on outside this subsystem.

4.5 Fields of Influence

I suggested above that a functional representation of causal powers fits well with the functional representations of force fields one finds in physics. One has to be careful when comparing powers with force fields, however. I have suggested above that powers be modelled as influence-valued functions on the set of possible stimuli. A force field, on the other hand, is a function on the set of actual space(-time) locations. So the two differ in two important ways. First, dispositions and force fields differ modally: dispositions deal with *possible* stimuli, while force fields deal with *actual* locations. In particular, force fields do not explicitly include the input modality that is characteristic of dispositional properties. Second, force fields are a function on the space of *locations*, while powers are a function on the space of *conditions*. Nonetheless, the relationship between causal powers and force fields is relatively straightforward.

Consider again electric charge. We will model the possession of charge Q with a function on the state-space of a two-particle system in which one particle, a, has charge Q and the charge and relative position of the second particle, b, are allowed to vary. Other properties like mass or spin are simply ignored. The function will associate each such state with a force (that is, an influence that the two particles exert on each other tending to accelerate them towards or away from each other). It turns out that if we fix the relative position of the two particles and allow only the charge of b to vary, then the force between the two will be proportional to the charge of b. This systematic relationship between the force and the charge of b means that we can factor out the effect of this charge. To do this we simply restrict the state-space still further by fixing the charge of b (to unit positive charge, say). We are then left with a two-particle state-space in which the only variable that can change is the relative position of the two particles. A function from such a limited state-space is equivalent to a function from a set of spatial locations. In particular, each location will be associated with a force of strength $\kappa_e Q/r^2$, where r is the distance between that location and a (and the force is directed towards an acceleration of the particles away from each other if Q is positive, and towards each other if Q is negative). That is to say, if we factor out the effect of changing the charge of b, the function representing the disposition Q is exactly the same as the function representing the electromagnetic field due to Q. Thus, we can regard the electromagnetic field at any point, x, due to a having Q as a description of the influence that a would exert if there were a unit positive charge located at x. The full description of Q—factoring back in changes in the charge of *b*—is regained when we note that the force exerted on an arbitrary charge, *q*, located at *x* is just *q* times the force exerted on a unit positive charge at *x*.

4.6 Probabilistic Power and Influence

One might worry that the discussion so far has assumed that powers are deterministic. The requirement that functions only associate one member of the range with a given element of the domain means that each stimulus will only be associated with a single influence. So, if you fix the stimulus, you fix the influence. For simplicity I will indeed be assuming this kind of determinism in most of my discussions in this book. However, there are at least two ways that we could build indeterminism into this model if we wanted. First, we could model powers as functions that associate each stimulus with a probability distribution over possibly manifested influences. In other words, given a particular stimulus, a power might manifest influence *A*, or *B*, or *C*, or ... with a determined probability for each. Second, we could build probabilities into the way that influences bring about effects. In this case, the influences acting on an entity determine only a probability distribution over possible changes in behaviour. The first way is naturally understood as modelling powers that are non-deterministic in their manifestations, while the second is naturally read as modelling influences themselves as non-deterministic. So the difference is whether the indeterminism is associated with the input modality or the output modality.

4.7 Conclusion

In this chapter I have argued that the causal powers that are assumed by the reductive method should be understood as dispositions to manifest causal influence. Such an understanding fits nicely with the reductive method, but it also allows for a simple conditional analysis of dispositionality, since the basic causal powers assumed by the method are immune to finks, antidotes, and mimics, at least within the range of conditions in which the reductive method is applied. I have also argued that many powers—even fundamental powers—are multi-track, which suggests that powers are best represented by functions rather than singular subjunctive conditionals. Along the way, I have argued that the 'dispositional modality' identified by Mumford and Anjum is not a new way to understand the modality that is characteristic of dispositions, but rather a new modality that is additional to that possessed by dispositions. The vectors in Mumford and Anjum's vector model of causation, I suggest, represent manifested influences, rather than dispositions. Thus, the dispositional modality associated with these vectors is actually a property of causal influence.

The result of these considerations is a new model of causal powers as influence-valued functions on a state-space. This model is appropriate for the reductive method of explanation, it deals well with counterexamples involving finks and antidotes, and it ties Mumford and Anjum's concept of dispositional modality to more traditional accounts of dispositionality. All in all, I suggest, this model of causal powers has a lot going for it.

5

Putting Things Together

The Assumptions of Reduction

In previous chapters I argued that reductive explanation presupposes the existence of causal powers that manifest causal influences. This chapter develops a model of reductive explanation which clarifies the role these powers and influences play. There are two general motivations for developing such a model. First, reductive explanation is intrinsically worthy of detailed investigation, since it plays such an important role in our attempts to understand the world. Second, a detailed model of the reductive approach to explanation allows us to identify a number of metaphysical presuppositions of the approach—'metaphysical', at least, in the sense that they bear on issues of traditional concern to metaphysicians. In so far as the model presented here is an accurate account of how reductive explanation works, the great success of the reductive approach will be evidence for the truth of these presuppositions, and so will give us an empirical handle on metaphysical questions. The presuppositions unearthed in this chapter will also inform the metaphysical picture that is developed in subsequent chapters.

The model of reductive explanation presented here involves four steps:

Step 1: Decompose the Complex System The first step is to identify the basic components of the complex system. A reductive explanation will attempt to explain or predict the behaviour of the complex system in terms of the behaviour of, and interactions between, these basic objects.

Step 2: Identify Relevant Powers and their Associated Subsystems The second step is to identify subsystems of basic objects that interact with each other. The interactions within these subsystems are assumed to be known (whether through observation or theory) and are assumed not to be affected by the fact that these subsystems are part of a larger whole.

Step 3: Calculate the Basic Influences The third step is to calculate all the relevant causal influences that act on the relevant basic objects as a result of their participation in the interacting subsystems identified in step two.

Step 4: Compose the Influences The fourth and final step is to figure out how each basic object will respond given all the influences that are acting on it. Once we know how each object responds, it is assumed we know everything we want to know about how the system as a whole responds.

These steps are discussed in detail in Sections 5.1–5.4 below. As mentioned in Chapter 1, we can distinguish an attempted reductive explanation from a reductive *method* of justifying or developing such an explanation. I will investigate the assumptions of this reductive method in Section 5.5.

This is the most technical chapter in the book and any readers who are averse to technicality should feel free to skip to Section 5.7. However, I suggest that you at least take some note of the definitions introduced along the way.

5.1 Decomposing the Complex System

A reductive explanation attempts to explain behaviours in a complex system in terms of the properties of, and interactions between, components of the system. In order to do this, we must identify, or posit, a relevant set of component entities and attribute causal powers to these entities. Let us call the set of components and causal powers referred to by a reductive explanation the *explanatory basis* for that explanation. In this section and the next, I will consider some of the features required by any explanatory basis that is to be used for reductive explanation. I begin, in this section, by looking at the set of component entities. In the next section, I will consider the causal powers that are attributed to these entities.

Let us begin by asking what kind of component entity can serve in the explanatory basis of a reductive explanation. Since reductive explanations occur in many sciences, and in everyday life, I take it that the component entities to which they refer could come from many different types. These entities could include, for example, stars, people, cars, rocks, bacteria, populations, molecules, atoms, particles, electromagnetic fields, and ghosts. However, there is some limitation on the kind of entity that can play the role of a component part in a reductive explanation. Consider what Stuart Glennan says about the parts that are involved in mechanistic explanations:

> Parts may be simple or complex in internal structure, they need not be spatially localizable, and they need not be describable in a purely physical vocabulary. In certain contexts, for instance, one might wish to consider genetic mechanisms whose parts are genes or information processing mechanisms whose parts are software modules or data structures. There are, however, certain kinds of entities which, to prevent [the concept of a mechanism] from being vacuous, should not be allowed to be parts of mechanisms. The parts of mechanisms must have a kind of robustness and reality apart from their place within that mechanism. It should

in principle be possible to take the part out of the mechanism and consider its properties in another context. Care must be taken so that parts are neither merely properties of the system as a whole nor artifacts of the descriptive vocabulary. I shall summarize these restrictions by saying that parts must be objects.

(Glennan, 1996, 53)

As described in Chapter 2, an essential feature of reductive explanation is that, when giving such an explanation, we consider parts in relative isolation. Thus, the robustness and reality that Glennan says are necessary for the parts referred to in mechanistic explanation are also necessary in reductive explanation. The parts must have an existence separate to their role in the larger system, such that the parts can be considered, at least in principle, in isolation from that system. In other words, the parts that are involved in reductive explanation must be *objects* in Glennan's sense.[1]

Glennan also points out that the component objects involved in a mechanistic explanation may have a complex internal structure, and the same is true in the case of reductive explanation. It follows that there will typically be numerous different ways to divide a complex system into components. A quantity of water for example can be regarded as composed of H_2O molecules, or composed of atoms, or composed of subatomic particles. A human population might be divided up into individuals, into the set of males and the set of females, into age groups, and so on.

Nonetheless, a reductive explanation involves a set of component objects that are treated as indivisible atomic objects for the purpose of explanation, even if they are in fact complex. For example, the explanation of the behaviour of falling water in Chapter 1 made reference to interactions between water molecules. For the purposes of the explanation, the molecules were treated as un-analysed atoms, even though they are actually complex objects.

Notice, however, that if we are to successfully treat a set of objects as indivisible atoms for the purpose of an explanation, then this set of objects must satisfy certain conditions. First of all, it should not be the case that one object in the set is composed from other objects in the set. If a is composed from b and c, then this will imply certain relations between these objects. But if we treat all three as atomic, then we will be blind to these relations and this could lead to problems. So, for example, suppose that b and c both exert some influence on a fourth object, d. This fact might mean that the composite object a exerts an influence on d, though

---

[1] Indeed, I am tempted to go further and suggest that 'reality apart from the system as a whole' is at least partly constitutive of our general concept of an object, and that our justification for dividing the world up into objects as we do has something to do with the role that these objects can play in reductive explanations. If this is right, then the considerations of this chapter may help with an answer to van Inwagen's special composition question: When is it true that there exists a concrete object y such that a collection of concrete objects composes y? (van Inwagen, 1990, 30) However, I will not follow up this suggestion here.

the influence that a exerts will be a composite of the influences exerted by b and c, and so will be nothing over and above those influences (see Chapter 6 for a discussion composite influences and whether they are something in addition to their components). But if we treat a as an atomic object, then we will be blind to the relation between the influence that it exerts and the influences exerted by b and c, and hence we may be led to mistakenly count the influence that a exerts in addition to influences exerted by b and c. Second, note that if objects in the set are allowed to be complex entities, then there is a possibility that two or more of these objects might share components. But if two objects in our set overlap in this way, then once again there will be relations between the two objects that we are blind to when we treat the objects as atomic. So, if we are to successfully treat a set of objects as atomic for the purpose of a reductive explanation, then these objects must be distinct, in the sense that no object in the set is a component of any other object in the set, and no object in the set overlaps with any other object in the set.

In any reductive explanation, therefore, the component entities in the explanatory basis must be *basic objects*, as defined below:

Definition 10 (Basic Objects). Given a complex system S, a set of *basic objects* of S is a set of entities such that:

1. Each entity is a component of S.

2. Each entity is an object in Glennan's sense (each entity is, in principle, capable of existing in isolation from S).

3. Each entity is distinct (no entity in the set is a component of, nor overlaps with, any other entity in the set).

Finally, if a set of basic objects is going to be adequate for a reductive explanation of some phenomenon, then the phenomenon in question must be related to the set of basic objects in an appropriate way. Consider again the explanation I provided in Chapter 1 for why a sheet of water breaks up as it falls. This explanation involved decomposing the sheet of water into molecules and describing the properties of, and interactions between, these molecules. The phenomena we wanted to explain, however, were why the water originally forms a sheet, and why this sheet then disintegrates. The behaviour of the collection of *individual* water molecules will only be useful in an explanation of why the water *as a whole* forms a sheet, if this property of the whole is somehow determined by the behaviour of the parts. This is indeed the case in this example. Whether the water as a whole forms a sheet or droplets is determined by whether the individual molecules are all arranged in a roughly two-dimensional surface with little distance between neighbours, or whether they are arranged in small separated clumps). In general, because reductive explanation seeks to explain properties of a complex system in terms of properties of its parts, the explanation assumes that the relevant properties of the whole are determined by the properties of the parts. If knowing the state of all the parts did not tell us what we want to know about the state of the whole, then the reductive approach would be useless.

This required relation between the basic objects and the phenomenon being explained can be conveniently stated using the concept of supervenience. A set of properties A supervenes upon another set B just in case no two things can differ with respect to A-properties without also differing with respect to their B-properties. In other words, the A-properties are determined by the B-properties. Our explanation for the behaviour of falling water, for example, assumed that the relevant properties of the water (forming a sheet, forming droplets) supervene on the state of the component water molecules.

Putting these thoughts together, we are now in a position to carefully state the first assumption of reductive explanation:

Assumption 1 (Supervenience). *When attempting a reductive explanation of some phenomenon P in complex system S, we assume that there is a set of basic objects of S such that the relevant properties of P supervene on the properties of, and relations between, these basic objects.*

The ultimate justification for a particular choice of basic objects in an explanatory basis is that the objects and their causal powers satisfy this assumption and others listed below, for it is only by satisfying these assumptions that the explanatory basis is adequate for reductive explanation. In particular, the basic objects must interact with each other in a consistent way. Suppose, for example, that you were attempting to explain the history of modern society in terms of the interaction between two classes, the bourgeoisie and the proletariat. If the interactions between these classes are not consistent in different situations, then we might attribute this inconsistency to interactions between smaller groups within these classes, and so be led to take these smaller groups as our basic objects. Alternatively, we might seek a reductive explanation in terms of basic objects that cut across the original choice. For example, we might seek to explain the history of modern society in terms of the interaction between male and female. Furthermore, different choices of basic objects may satisfy the assumptions of reductive explanation to a greater or lesser degree, and the greater the degree of satisfaction, the better the explanation will be. In particular, the greater the degree to which a choice of basic objects satisfies the assumptions, the more useful this choice will be in explaining multiple phenomena.

5.2 Identifying Powers and their Associated Subsystems

Once we have chosen a set of basic objects, a reductive explanation proceeds by attributing causal powers to these objects and then explaining the behaviour of the complex system in terms of the influences manifested by those powers. So let us now turn our attention to the properties that are required of a set of causal powers if these powers are to serve in the explanatory basis of a successful reductive explanation.

Reductive explanation must begin with a set of basic objects, and a set of basic causal powers. However, I will argue in Chapter 6 that just as there can be composite objects that are composed of more basic objects, there can be composite powers that are composed from more basic powers. Put simply, a power P will be composed from powers Q and R if the influences that P manifests are themselves composed from influences that are manifested by Q and R. I will also argue in Chapter 6 that a composite influence is nothing over and above the component influences from which it is composed. The possibility of composite powers has similar implications for reductive explanation as did the possibility of composite objects.

First, just as we allowed composite objects into our explanatory basis, we can allow composite powers so long as they are treated as simple, atomic powers for the purpose of the explanation. So, for example, it is often convenient to treat the Earth as a single basic object with a single basic gravitational power. This is so even though the Earth is in fact composed of many objects and the gravitational power of the Earth is composed from the gravitational powers contributed by these objects.

Second, just as the basic objects in our explanatory basis must be distinct, so too must be the causal powers in the basis. Suppose, for example, that our explanatory basis includes the gravitational power of each atom that makes up the Earth, then we do not want to include, in addition, the composite gravitational power of the Earth as a whole. For if we were to treat the gravitational power of the Earth as a simple, non-composite power then we will be blind to the fact that the influences it produces are nothing over and above influences manifested by its component powers. This blindness will lead us to overcount the influences at play.

Finally, a reductive explanation attempts to explain some phenomenon by reference only to the basic powers of a set of basic objects. Thus, in order for a set of basic powers to be adequate for a reductive explanation of some phenomenon, this set of powers must be complete, in the sense that there are no influences relevant to the phenomenon that are not manifestations of the powers in the set.

In any reductive explanation, therefore, the powers in the explanatory basis must be *basic powers*, as defined below:

Definition 11 (Basic Powers). Given some phenomenon P, a set of *basic powers* for explaining P is a set of causal powers such that:

1. Each power is distinct (no power in the set is a component of, nor overlaps with, any other power in the set).

2. All influences that are relevant for the phenomenon P are manifestations of powers in the set.

We will say that an influence is basic for the purpose of an explanation if it is the manifestation of a basic power in the explanatory basis used in that explanation. Since the set of powers in an explanatory basis for some phenomenon is taken to be complete, if we wish to take account of all the influences relevant for that phenomenon, we need only take account of all the basic influences that are manifested.

Now, whether, and how, a power is manifest can depend on the situation it is in, and this situation can involve more than one basic object. For example, the force exerted by one charged particle on another will depend on the strength of the charges and the distance between the two particles. Thus, to figure out the force on one of these particles we need to consider the state of the *two-particle* system. If we know the state of only one of these particles, then we cannot calculate the force either exerts on the other. On the other hand, once we know the charge of the two particles, and the distance between them, then we can figure out the force each exerts on the other, since this force does not depend on any further facts about the environment in which the two-particle system is placed. In this sense the force each particle exerts on the other due to their charges is invariant given the state of the two particle system. The systems-in-isolation character of reductive explanation assumes that *all* the relevant powers are invariant in this way. That is, each power is associated with a subsystem (or, more accurately, a *type* of subsystem—see below) such that the influence the power manifests is invariant given the state of that subsystem.

As discussed in Chapter 4, the invariance of a causal power is straightforwardly captured when we use an influence-valued function to represent the causal power. The domain of such a function will be the state-space of some system, and the function will associate a single influence (or single probability distribution over the space of influences) to each possible state of the system regardless of what else is going on in the world. In particular, the function does not take note of whether the system is a subsystem of some larger system.

Note that the domain of the function which represents a power is the state-space of the associated system, not the system itself. Thus, the function can be applied to any system which has the same state-space. Since the state-space is the space of all possible values for all the relevant properties of the system, what this means in practice is that the function representing a causal power can be applied to any system of this same type. That is, if a power is associated with a particular system, S, then the function representing that power can be applied to any other system that instantiates the relevant properties of S. So everywhere the power is instantiated, it can be represented by the same function operating on the same state-space.

In order to give a reductive explanation in some complex system S, then, it is not enough to divide S into basic objects: we must also identify the subsystems of S that are relevant to the powers that these basic objects possess. Each of these subsystems will be constituted by one or more of the basic objects.

Suppose, for example, that we are giving a reductive explanation in some system S_{abc} that is constituted by three basic objects a, b, and c. Suppose, further, that each of these objects has some positive mass, and that a and b, but not c, have some electric charge. In order to calculate the influences that the causal powers of mass and electric charge manifest in this complex system, we need to identify the subsystems that are relevant to these powers. According to Newton's theory of universal gravitation (which, for simplicity, we will assume is correct), gravitational influences occur between any pair of massive objects and are a function of the masses of the two objects and the distance between them. So the function representing gravitational power will apply to all subsystems consisting of two massive objects separated by some spatial distance. There are three such subsystems in S_{abc}: S_{ab}, which consists of the objects a and b; S_{ac}, which consists of the objects a and c; and S_{bc}, which consists of objects b and c. Similarly, according to Coulomb's law, there will be forces (that is, influences) between any two charged objects, and the forces are a function of the charges of the two objects and the distance between them. So the function representing electric charge will apply to all subsystems consisting of pairs of charged objects. Since we have supposed that c is not charged, there is only one subsystem of S_{abc} relevant to the power of electric charge: S_{ab}. Figure 5.1 shows how our complex system is decomposed.

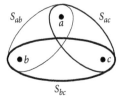
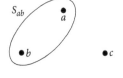

(a) Subsystems relevant for gravitational powers

(b) Subsystems relevant for electromagnetic powers

Figure 5.1. Identifying relevant subsystems

So far, nothing I have said about the basic powers in an explanation connects them to the system whose behaviour is being explained, nor to the set of basic objects that has been chosen for the explanation. However, if a set of basic objects and a set of basic powers are to work together as the explanatory basis for some phenomenon, then there must be some kind of connection between these basic powers and the basic objects. Intuitively, we want the basic powers to be the powers possessed by the basic objects, but putting things this way is not particularly helpful. What exactly does it mean for a power to be possessed by an object, and what implications does this have for reductive explanation? I believe that we can give a much clearer account of the connection between the basic objects and powers in an explanatory basis by referring to the subsystems associated with these powers, as discussed above. This thought motivates the following definition of an explanatory basis:

Definition 12 (Explanatory Basis). An *explanatory basis* for some phenomenon P in complex system S is a set of basic objects of S together with a set of basic powers for P such that each power in the set can be represented by a function whose domain is the state-space of some subsystem of basic objects of S.

Here, the connection between the basic powers and the basic objects in an explanation is ensured by the fact that the subsystem associated with each basic power is formed from the basic objects.

One might worry, at this stage, that if I am restricting reductive explanation to make use only of influences manifested by the basic powers of the basic objects that make up some system, then, by definition, reductive explanations cannot incorporate the possibility of influences from the environment of S. But this problem is easily remedied: we simply allow that the environment can be included as another basic object with its own basic powers. There is no need to alter our definitions to allow this, since we can stipulate that any case where the environment needs to be taken into account is a situation in which the environment is part of the system S. The intuitive separation between system and environment is reflected in the fact that the environment is always treated as a basic object.

Let us now turn our attention to an interesting complication. One thing that Figure 5.1 makes clear is that the subsystems used in a reductive explanation can overlap. In the example above, there are influences exerted on a by both b and c. Thus, to figure out all the influences felt by a, we need to consider its role in both S_{ab} and S_{ac}. Both systems can contribute an influence on a which will need to be taken into account. Notice however, that object a is also part of two other subsystems, namely S_a, which is the system consisting in a alone, and S_{abc}, which is the system consisting of all three objects. This raises an interesting question: must causal powers only be associated with subsystems of two basic objects, as in the example above, or can powers be associated with subsystems of any number of objects?

Let us define the rank of a power as follows:

Definition 13 (Rank of Power). A power P to manifest an influence on a is of *rank n* (relative to a given set of basic objects) iff n is the cardinality of the smallest system of basic objects with respect to which the influence of P on a is invariant.

Our question is then whether there can be any powers with a rank other than two. The simple answer is yes.

Recall the radium atom discussed in Chapter 4. If Lowe is right that the decay of such an atom requires no stimulus at all, then the atom's probabilistic power to exert an influence directed toward decay is a power that is triggered by simply being a radium atom. Thus, the triggering of the power involves only the radium atom itself, and so the power is of rank one. Consider also the example above in which b and c each exert an influence on the acceleration of a. Call these two influences I_{ab} and I_{ac}. Now, I argue in Chapter 6 that the combination of two influences that are directed at a change in the same property of a single object is itself an influence (the resultant force in this case). Call the resultant influence in this case I_{abc}. In order to calculate I_{abc}, we need to know the properties of and relations between all three objects, a, b, and c. So I_{abc} will not be invariant given either S_{ab} or S_{ac}, but will be invariant given S_{abc}. In other words I_{abc} is the manifestation of a third-rank power.

Now if our explanatory basis includes the second-rank power to manifest I_{ab} and the second-rank power to manifest I_{ac}, then it cannot also include the third-rank power to manifest I_{abc}, since I_{abc} is not distinct from I_{ab} and I_{ac} in the appropriate way. This is not a problem, because the third-rank power is a composite and can be safely ignored if we have already taken into account its component powers. But there is nothing in principle to stop us taking this third-rank power as a basic power in our explanatory basis instead of the two second-rank powers. In general, the explanatory basis for some reductive explanation may include basic powers of any rank.

All else being equal, however, lower-rank powers are preferable in an explanatory basis. This is because small systems of a given type are likely to be instantiated much more often than a more complex system, and so an explanatory basis with lower-rank powers will be useful in more cases than a basis with higher-rank powers. The system shown in Figure 5.1, for example has three instantiations of systems involving two charged objects, and only one instantiation of a system involving three charged objects. Nonetheless, there is nothing incoherent about allowing higher-rank powers into a basis, and there may be situations in which it is desirable, or even necessary, to do so (see Chapter 10 for further discussion of high-rank powers).

The possibility of basic higher-rank powers requires another assumption for reductive explanation. Suppose that a complex system S is composed from n basic objects. Then, there will be subsystems of all sizes from 1 to n (the subsystem of size n is just S itself). In principle, any of these subsystems could be associated with a power that is not a composite of lower-rank powers. To be sure that we have taken account of all basic influences in S, then, we will need to have studied all of the subsystems of S. In practice, of course, we do not do this. The great power of reductive explanation comes from the fact that it explains the behaviour of complex systems in terms of relatively low-rank (e.g. second-rank) interactions. For systems of any great complexity, it would be very burdensome to have to consider all the possible higher-rank powers. More importantly, perhaps, the higher the rank of the powers that need to be considered, the more specific these powers are to the particular system being studied and hence the less general and less unifying the explanation becomes. To see the most extreme possibility, suppose that the system S has n basic objects, and suppose that S has a power of rank n. Then, to take account of this power we will need to study the entire system as a whole, and cannot rely on our knowledge of the interactions between smaller parts. The power and practicability of reductive explanation, therefore, relies on us limiting our attention to powers of low rank. With this in mind, let us define the rank of a reductive explanation as follows:

Definition 14. A reductive explanation is of *rank N*, iff *N* is the rank of the highest-ranked causal power in the explanatory basis being used.

In general, then, when giving a reductive explanation in system *S*, we must pick some number *N*, which is less than the number of basic objects in *S*, and then look for an explanation of rank *N*. When looking for an explanation of rank *N*, however, we must assume that we can ignore any basic powers in *S* that have a rank greater than *N*. In other words:

Assumption 2 (Limited rank). *When attempting a reductive explanation of rank N of some phenomenon P in complex system S, we assume that there is an explanatory basis for P in S that does not include any basic powers that have a rank greater than N.*

5.3 Calculating the Basic Influences

Once we have chosen an explanatory basis and identified the subsystems that are associated with the basic powers, the next step is to calculate all the basic influences generated within the system that are relevant to the phenomenon we wish to explain. This step is fairly straightforward, since each basic influence is manifested by a single basic power, and the influence that each power manifests is determined by the state of the subsystem associated with that power. Typically, we will have a reasonably good understanding of these causal powers, and so we will be able to calculate the influences generated within each subsystem. This is where the reductive approach gets its great power; we can calculate these influences even though the subsystems that generate them may be in a larger context that we have never before encountered.

Suppose, for example, that we wished to explain, or predict, the behaviour of object a in the three object system that was depicted in Figure 5.1. All the relevant powers in this system are second-rank, so the basic influences acting on a will come from the two two-object subsystems of which a is a part, namely S_{ab} and S_{ac}. The influences on a coming from S_{ab} are the influences that a feels due to its interaction with b, and there will be two of these, one electromagnetic and one gravitational. Call these I_{ab}^e and I_{ab}^g respectively. a will also feel one gravitational influence due to its relation to c. Call this I_{ac}^g. Thus, there are a total of three basic influences acting on a. This situation is illustrated in Figure 5.2.

I have spoken here of 'influences coming from a two-object subsystem', rather than speaking of 'the influence of one object on another'. There are

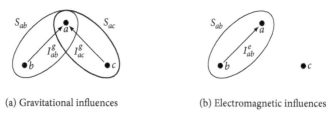

(a) Gravitational influences   (b) Electromagnetic influences

Figure 5.2. Calculating the influences on a

interesting questions here as to whether these two ways of speaking indicate different ways the world might be, and, if so, which is the correct description? It is tempting, for example, to regard the power that manifests influence I^g_{ab} as a power that is possessed by the system S_{ab} as a whole, rather than being possessed by either or both of the objects a and b. If we view the power this way, this might support the view that I^g_{ab} 'comes from' S_{ab} rather than 'coming from' b. On the other hand, the gravitational mass of a is something that remains constant in every two-object system in which a is a part. Thus, it would make sense to regard a alone as the possessor of this causal power, and regard gravitational force as being exerted by one object on another. I am not sure exactly which view here is correct, or even if they really are distinct views.[2] For the purpose of this book I will simply overlook any possible differences here. I do not believe that this decision is important for any of the conclusions I draw here.

However, if influences can be said to come *from* a two-object system, then could they not act *on* a two-object system? Why should we think that the influences produced by these powers act on individual components of the system rather than the system as a whole? Why, for instance, do I say that I^g_{ab} is a force acting on a, rather than a force acting on the system S_{ab}? The answer is that influences must act on basic objects individually rather than on subsystems of objects. The reason for this is that subsystems can overlap. Suppose, for example, that there is an influence on the system S_{ab} and another influence on the system S_{ac}. Since the two influences are not acting on completely disjoint systems, the behaviour that occurs must be the result of the two influences acting in combination (the behaviour of a at least, will be affected by both influences). On the other hand, since the two influences act on different systems, there does not seem to be any way to make sense of their combination. To put the problem another way, the behaviour of a can be affected by all the influences that act on systems of which a is a member. Thus, a can be affected by both the influence acting on S_{ab} and the influence acting on S_{ac} and we must therefore add these two influences to determine a's behaviour. However, b is not part of both systems and so its behaviour is only determined by

---

[2] The issue here seems closely related to Martin's (2008) claim that many causal powers require 'mutual manifestation partners'.

one of the influences. Hence the combination of influences coming from S_{ab} and S_{ac} cannot act on S_{ab}, since it affects a and not b. In order to combine coherently, then, influences must act on individual non-overlapping components. That is to say, influences must act on basic objects. Fortunately, Assumption 1 ensures that the phenomenon we wish to explain is determined by the behaviour of the basic objects. This means that we will be able to use information about influences acting on basic objects to explain what we want to explain.

5.4 Composing Influences

Once we have figured out all of the basic influences that are acting on each object, the final step is to figure out the combined effect of all these influences acting at the same time. According to Assumption 1, any phenomenon that we wish to explain or predict using the reductive approach must supervene on the state of the basic objects in our explanatory basis. Thus, we can proceed by dealing with the basic objects one at a time. Considering each basic object in turn, we calculate the combined effect of the influences acting on that object.

When proceeding in this way, we are obviously assuming that all relevant changes in each basic object are determined by basic influences and not by something else. But we must also assume that only some basic influences are relevant—there are very many, possibly infinitely many, influences in the universe and it is simply not possible to take account of them all. Typically, for example, reductive explanation assumes that the only influences that are relevant to changes in some basic object a are influences that are themselves directed towards changes in a. While it is logically possible that a force acting on b may combine with a force acting on c to produce an acceleration in object a, the theory of forces rules out such a possibility. Similarly, we typically assume that the only influences relevant to a change in a at time t are influences that are acting on a at time t. These typical assumptions are possibly related to assumptions about spatial and temporal locality, though I am not yet clear about the nature of this relation. Perhaps assumptions of spatial and temporal locality will imply or motivate these assumptions about relevant influences, but the implication or motivation could also go the other way.

A further filter that we use to narrow the field of relevant influences is that, when considering changes in some property P (of object a at time t), we assume that the only influences that are relevant are influences that are directed towards changes in P (in a at t). Again, it is logically possible that this assumption is false. It is logically possible, for example, that two forces acting in combination might bring about a change in colour, or electric charge, rather than a change in acceleration.

We also typically ignore influences that we believe to be too weak to be significant, but this seems to be a matter of practicality, while the assumptions above seem to be matters of principle. We ignore weak influences whilst admitting that they do make some small contribution to the properties we are interested in. But when we ignore influences that act on other objects, times, or properties than those we are interested in, we assume that the influences we ignore make no contribution whatsoever to the changes we are concerned with.

We can sum up the assumptions discussed in the last three paragraphs as follows:

Assumption 3 (Changes Determined by Influence). *The changes in property P of some basic object a, at time t, are determined by the set of all and only the relevant basic influences. Typically we assume that influences are relevant if and only if they are individually directed towards changes in P of a at t.*[3]

Three issues rate mention at this point. First, by 'changes in the behaviour of a basic object', I mean changes compared to the behaviour that object would exhibit in isolation. The motion of a particle under the influence of a force, for example, will depend on the particle's initial velocity as well as the force. The effect of the force is thus best understood as causing a change to the motion of the particle rather than causing the motion. Second, this assumption could easily be generalized to include probabilistic cases; we simply assume that the influences acting on an object determine only a probability distribution over possible changes in behaviour. Finally, we do not actually need to include this as one of the basic assumptions of reductive explanation, since it is implied by Assumption 4, which I will set out below.

5.5 Assumptions of the Reductive Method

Assumptions 1–3 lay out what is required for a reductive explanation to be possible—they are assumed in any attempt at reductive explanation. In the remainder of this chapter, I wish to investigate the assumptions required for the reductive *method* for developing and/or justifying a reductive explanation, whereby we attempt to provide a reductive explanation of some phenomenon in a complex system by studying the components in some other situation (e.g. isolation) and use what we learn to explain their behaviour in the complex system.

5.5.1 The Algebra of Influences

In order to calculate how a property of some object changes at a given time, we need to combine all the influences that are directed at a change in that property

---

[3] I am here assuming that *P* is not a disjunctive property.

in that object at that time. How is this done? Given any set of relevant influences acting on an object, there will be a function which takes these influences as an input and outputs the changes in the behaviour of that object. Call this the *total influence function* for these influences. So if we know the total influence function for the set of relevant influences, then we will have no problem figuring out the result of these influences acting in combination.

Now, recall that the reductive method takes something (like, say, the total influence function) that we have learned in one situation and applies it in another. But the total influence function is defined relative to a set of relevant influences, and there is no guarantee that the function for one set of influences will be anything like the function for some other set. So, unless we make further assumptions, there is no reason to think that learning the total influence function in one case will tell us anything about how to compose influences in another case.

As discussed above, the set of influences that are relevant to a change in property P of object a at time t is typically taken to be the set of influences that are individually directed towards changes in P of a at t. Now, we typically assume that the world is temporally homogeneous, in the sense that the way things interact with each other does not depend on the particular time. If this is the case, then it would follow that, for any times t_1 and t_2, the total influence function for a set of influences acting on P of a at t_1 will have the same form as the function for a similar set of influences acting on P of a at t_2. Similarly, we typically assume that the ways that objects interact is determined by the properties that these objects have rather than some individual essence, or haecceity, of the objects. Thus, we assume that, for any objects a and b that have the same relevant properties, the total influence function for a set of influences acting on P of a at t_1 will have the same form as a similar set of influences acting on P of b at t_2. In other words, we assume that the total influence function for a set of influences is not determined by the time at which they are acting, nor the object that they are acting on, but rather the property that they are directed towards.

Let us say, then, that two influences are of the *same type* if they are directed towards changes in the same property. So, for example, all forces are of the same type since they are directed towards changes in acceleration. Then our assumption is that once we learn the total influence function for a given type of influence, then we can apply this to influences of that type acting on any object at any time.

But how do we learn about the total influence function for a given type of influence? The power of the reductive method is that it can be applied to complex systems that are not already well understood. In such situations, we will not have previously observed the behaviour of the objects under the combined effect of all the influences they feel in this complex system. Thus, when applying the reductive method to new cases, we cannot expect to know the total influence function from direct observation of the effect of the relevant set of influences. Instead, we must figure it out based on our knowledge of the component subsystems. In particular, we need to be able to take our knowledge of how influences act together in simple cases and apply this to the complex case.

What we need is a set of laws for composing influences, and we need to be able to learn these laws from simple cases and then apply them in complex cases. If this method is to work, then influences are not the only thing that must remain invariant when moving from subsystems to more complex systems: the composition laws must also. Suppose, for example, that we have learned how two influences of a certain kind act in combination by studying simple systems in which only two influences of that kind are at play, and suppose that we then turn our attention to a more complex situation in which there are three or more of these influences. Conceivably, the function which describes how three such influences combine could have nothing at all to do with the function which describes how two such influences combine. But if this were the case, then we could not use the knowledge we gained by studying the simple system to understand the complex one. The reductive method, therefore, works by assuming that the composition function for influences remains the same, in some sense, from cases in which we have two influences to cases where there are three or more influences.

But in what sense could the function be 'the same' in the three-influence case as it is in the two-influence case? The two-influence case involves a two-place function, while a three-influence case involves a three-place function, so it cannot be the same function in each case. The solution to this problem is to insist that influences must be composed pairwise. To compose three influences u, v, and w, for instance, we can first compose u and v using the rule learnt in the two-influence case, and then take the result of that and compose it with w using the same two-place function. In this way we can compose an arbitrary number of basic influences using a two-influence composition rule.

Now we run into another problem. By studying the composition of two influences in the simple system, we come up with a function which relates a pair of influences to the observable behaviour that is produced when those two influences act together. But behaviours are not themselves influences (since, as I argued in Chapter 3, behaviours are not invariant in the way that influences are). Thus, we cannot use the output of such a function as an input to that rule, and so we cannot use such a rule to compose three or more influences pairwise.

This problem is also easily solved. We simply introduce the notion of a *resultant influence*. Rather than constructing a rule which relates each pair of influences to the effect they produce in combination, we construct a rule which relates each pair of influences to a third, resultant influence. The resultant of two influences is defined to be an influence that is directed towards the effect that would be produced by the pair in combination when no further influences are at play. In other words, if u and v are any two influences we can define their composition $u \oplus v = r$, where the resultant r is the influence that is directed to the outcome produced by u and v when acting together in the absence of any other influences.[4]

We can now use this composition rule to calculate the resultant of any number of influences. To calculate the joint effect of u, v, and w, for example, we simply calculate the resultant $(u \oplus v) \oplus w$. The resultant of four influences will be $((u \oplus v) \oplus w) \oplus x$, and so on. Note that this trick cannot be pulled using a composition operator that takes more than two influences. For example, suppose we learn that three influences acting together compose according to a three-place operator. There is no consistent way to string together a three-place operator to produce a four-place operator, a five-place operator, and so on. The problem just gets worse if we start with operators of even higher arity.

The assumption that the laws of composition remain invariant when we move from systems with two influences to systems with higher numbers of influence, then, comes down to the the assumption that influences compose pairwise, so that the result of $u_1, u_2, \ldots$ and u_n acting together is the influence $(\ldots (u_1 \oplus u_2) \oplus \ldots) \oplus u_n$. In Chapter 10, I consider an example of a reductive explanation in which this assumption fails, meaning that the explanation cannot be derived or justified by the reductive method.

Let I_{Pa} be the set of all possible influences of type P that could act on some object a. Then the discussion above shows that influence composition defines a binary operator $\oplus$ on I_{Pa}. This operator has some interesting mathematical properties.

First, Assumption 3 tells us that the outcome that occurs when two influences of type P act together on object a is a change in the property P of a. So the resultant of any two influences in I_{Pa} will be another influence in I_{Pa}. In other words I_{Pa} is closed under the operation of $\oplus$:

Closure For all $u, v \in I_{Pa}$, $u \oplus v \in I_{Pa}$.

Second, notice that the order of composition does not represent anything in reality. When calculating the result of two influences acting together, we are interested in cases in which the two influences exist *at the same time*, so we are not modelling a situation in which one occurs and then another. Nor is there any non-temporal priority that should be represented by the order of composition. Thus, the composition of influences has the following two features:

Commutativity For all $u, v \in I_{Pa}$, $u \oplus v = v \oplus u$

and

Associativity For all $u, v, w \in I_{Pa}$, $(u \oplus v) \oplus w = u \oplus (v \oplus w)$.

---

[4] I have assumed that resultant influences so defined are indeed influences. This view will be defended in Chapter 6.

Closure, Commutativity, and Associativity mean that $(I_{Pa}, \oplus)$ is what mathematicians call a *commutative semigroup*.

Putting the lessons of this subsection together, then, we see that the reductive method makes the following assumption about the nature of influence composition:

Assumption 4 (Algebra of Composition). *Let I_{Pa} be the set of influences of type P acting on object a. There exists a two-place operator $\oplus$ such that:*

1. *The result of influences $u_1, u_2, \ldots$ and $u_n \in I_{Pa}$ acting together at the same time, and in the absence of any other influences in I_{Pa}, is the outcome that would occur if the influence $(\ldots (u_1 \oplus u_2) \oplus \ldots) \oplus u_n$ acted on its own;*

2. *$(I_{Pa}, \oplus)$ is a commutative semigroup.*

Note that in the case of a single influence acting on its own (so $n = 1$ in the first clause above), Assumption 4 reduces to Assumption 3, and so, as mentioned above, Assumption 4 implies Assumption 3.

Semigroups are well-studied structures in the field of abstract algebra, but mathematicians in this field are often more interested in the richer algebraic structure of what they call a *group*. The set of influences together with the composition operator will form a group if it has two further properties: (1) there is a zero influence, which makes no difference to the outcome when composed with any other influence; and (2) given any influence there is an inverse influence that would cancel out the first to produce the zero influence if the two were acting together. We can formalize these two properties as follows:

Zero There is an influence in I_{Pa}—call it 0—such that for all $u \in I_{Pa}$, $u \oplus 0 = 0 \oplus u = u$.

Inverses For all influences $u \in I_{Pa}$, there is an influence, $-u \in I_{Pa}$, such that $u \oplus -u = 0$.

So, does the set of all possible influences contain inverses and a zero? We might be tempted to reason that it does, as follows: suppose, for the sake of reductio, that $(I_{Pa}, \oplus)$ does not contain a zero. Then we can add a new element 0 which is simply defined to be such that, given any $u \in I_{Pa}$, $u \oplus 0 = u$. Similarly, if $(I_{Pa}, \oplus)$ is missing inverses for some elements of I_{Pa}, then we can simply add these inverses.[5] So, we might say, if $(I_{Pa}, \oplus)$ is missing a zero or inverses, then we can add new influences to I_{Pa} to form a set $I_{Pa}*$ such that $I_{Pa} \subset I_{Pa}*$ and $(I_{Pa}*, \oplus)$ is a group.

---

[5] If u is missing an inverse, for example, then we can simply add an element, $-u$, which is defined to be such that $u \oplus -u = 0$. The new element will automatically have u as its inverse since $\oplus$ is commutative.

But, by assumption, I_{Pa} is the set of *all possible* influences of type P that could act on a, and so we cannot make I_{Pa} bigger by adding further influences. In other words it is not possible that $I_{Pa} \subset I_{Pa}*$. Thus, our assumption that $(I_{Pa}, \oplus)$ is missing a zero or some inverses leads to a contradiction, and hence the assumption is false.

The problem with this reasoning is that there is no guarantee that the new elements that were added to I_{Pa} in the reductio above are themselves influences. These new elements were defined purely formally and the definition did not include the requirement that they be directed towards some change in the property P had by a. In general, I can see no reason to assume that a set of influences will always contain a zero, nor any reason to think that such a set will always contain inverses for every influence. Thus, we cannot assume that influences form a group under composition.

Nonetheless, many practical implementations of the reductive method do assume the existence of a zero and of inverses. Recall that the reductive method depends on studying the behaviour of subsystems in isolation. How, in practice, can we isolate a system from any external influences? If the strength of an influence decreases with distance, then we can effectively isolate a system by moving it so far from other objects that any external influences are negligible. But another way that we can isolate a system is by shielding it from its environment. We can shield a system from external electromagnetic influences, for example, by placing the system inside a hollow conductor. External electromagnetic influences produce currents in the conductor, and these currents produce influences which precisely counteract the external influences in the region inside the conductor. In general, when we isolate a system by shielding it, we are attempting to construct a situation in which the external influences add to the influences of the shield in such a way that together they make no difference to the influences being studied in the isolated system. This kind of shielding, therefore, assumes the possibility of creating inverse influences which add to external influences to produce a zero resultant influence.

A third way that a system can be treated as isolated is when the external influences average out to zero. When studying the behaviour of a pendulum in the gravitational field of the Earth, for example, we typically neglect any electromagnetic interactions between the pendulum and the Earth despite the fact that we believe both the pendulum and the Earth to be made of electrically charged particles, and we believe electromagnetic forces to be much stronger than gravitational forces. The reason we can neglect these electromagnetic forces is that the basic forces produced by a very large number of randomly distributed positive and negative charges will almost certainly produce a resultant very close to zero. As with shielding, the possibility of this kind of isolation presupposes the possibility of counteracting influences, and thus the existence of inverse influences and a zero influence (or, at least, close approximations to them).

5.5.2 The Algebra of Effects

Having calculated the resultant influence on each basic object, we now need to know how the objects will behave as a result of feeling this influence. First, changes to the basic objects are calculated from the influences acting on them. Second, since the state of the whole system supervenes on the state of the basic objects (Assumption 1), these changes in the basic objects are used to determine the overall change to the whole complex system.

At first, this stage seems fairly straightforward. Each influence is characterized by the effect that it will bring about if acting in isolation, and the total resultant influence acting on an object is, by definition, acting in isolation. Once we have calculated the total influence acting on an object, therefore, we know that the change which occurs to that object will be the change that is characteristic of that total influence. The problem is that we only know the resultant influence as a sum of basic influences, and if we have not already examined this sum in the past, then we will not know what effect is characteristic of this resultant. So, for example, monoamine oxidase inhibitors (MAOIs) are a class of drugs that are sometimes prescribed to treat depression. Tryptophan is a dietary supplement that is also sometimes taken as an antidepressant. If the two drugs are taken together, however, the combination can result in headaches, confusion, hallucinations, coma, and even death. Prior knowledge of the mood-lifting influence of these drugs acting on their own is clearly not enough for us to predict the dangerous influence they will have in combination. In order to apply the reductive method, therefore, there needs to be some kind of systematic relation between the structure of influences under composition and the structure of their characteristic effects, such that knowledge of the characteristic effects of two influences can be used to learn something about the characteristic effect of their resultant. Let us, then, turn our attention to the nature of this relationship.

Given some basic object a, let Δ_a be the set of all possible changes to a. Now, each possible influence on a has a characteristic effect on a when acting in isolation. So, there is a function from I_{P_a} to Δ_a that maps the set of all possible influences on a to the set of possible changes to a. Call this function φ. So, for example, if f is a force acting on an object of mass m, then $\varphi(f)$ is an acceleration of the object in the direction given by f and magnitude given by f/m. Our task is to figure out what is required in order to calculate $\varphi(u \oplus v)$ from our knowledge of $\varphi(u)$ and $\varphi(v)$.

Now, calculating $\varphi(u \oplus v)$ from $\varphi(u)$ and $\varphi(v)$ means being able to compute a function g which takes two changes from Δ_a and returns another change from Δ_a such that $\varphi(u \oplus v) = g(\varphi(u), \varphi(v))$. The function g is a binary operator on Δ_a, so we can write it using the symbol $\star$ thus: $g(x, y) = x \star y$. What is required, then, is that $\varphi(u \oplus v) = \varphi(u) \star \varphi(v)$. Written this way, it is easy to see that the requirement is that the operator $\star$ on the space of changes has the same algebraic structure as the operator $\oplus$ on the space of influences, and these structures are coordinated by the function φ. In mathematical jargon, what is required is that φ is a *homomorphism* from $(I_{Pa}, \oplus)$ to $(\Delta_a, \star)$.[6] For example, the group of forces under vector addition is homomorphic to the group of accelerations under vector addition (so both $\oplus$ and $\star$ are the operations of vector addition).

The existence of an operator $\star$ that satisfies these conditions is guaranteed if, and only if, φ is one-to-one, meaning that no two elements of I_{Pa} are mapped to the same element of Δ_a. In other words, our ability to calculate the effect of combined influences (without having observed these combinations previously) relies on the following assumption:

Assumption 5 (Influence Characterized by Effect). *Causal influences are uniquely characterized by the effect that they would bring about if acting in isolation.*

This makes explicit something that I have been assuming in much of my discussion to this point, but note that it is not trivial. Up to this point there have really been two distinct ways of identifying influences: we can identify them via the effects they produce in isolation, and we can identify them by the way they combine with other influences (or, more precisely, by their place in the algebraic structure of the semigroup of influences). This assumption ensures that influences classified as the same in one of these respects will also be classified as the same in the other. In particular, it rules out situations in which two influences have the same effect when acting in isolation, but make different contributions when each is combined with some third influence.

Note that, given any logically possible change in a, there is a logically possible influence that could produce that change if acting in isolation. So, if I_{Pa} is the set of all logically possible influences on a, and Δ_a is the set of all logically possible changes to a, then every element of Δ_a will be the change that is characteristic of some element of I_{Pa}. In mathematical terms φ will not only be one-to-one: it will also be onto, and hence will be an isomorphism.

In Newtonian mechanics, the set of possible changes Δ_a is the set of possible accelerations, and this set inherits a vector-space structure from the space of possible spatial positions. It turns out that the composition operator, $\star$, is simply vector addition in this structure, and so the group $(\Delta_a, \star, e)$ is just the group associated with that vector-space structure. The isomorphism between $(I_{Pa}, \oplus, 0)$ and $(\Delta_a, \star, e)$ endows the set of forces with its own vector-space structure, and the composition of forces is also vector addition. This close relationship between forces and accelerations makes forces very easy to work with, but it also obscures some distinctions. For example, it is common to imagine forces to be directed spatially. Textbooks often speak of forces directed downwards or north-east. But if forces are influences as I have described them, then forces are not directed in physical

---

[6] If $(I_{Pa}, \oplus)$ has a zero, 0, then we must also require that $\varphi(0)$ is the zero of $(\Delta_a, \star)$.

space, but in some more abstract space of possible outcomes—in this case the space of possible accelerations. If we were being precise, then, rather than saying a force is directed downwards we should say that it is directed towards a downwards acceleration.

In general the relation between $\oplus$ and $\star$ is not necessarily so straightforward as it is in the case of forces. Suppose, for example, that we are concerned with influences whose characteristic effects can be represented by a single real number (examples might be changes to mass, or changes to temperature, or accelerations in one dimension). The set of real numbers together with the operation of addition forms a commutative group, and this group could be the one that is isomorphic to the group of influences (as would be the case if we were considering one-dimensional accelerations). But there are other possibilities. The non-zero real numbers under the operation of multiplication also form a commutative group (zero must be left out, as it does not have an inverse). If this were the group that is isomorphic to the group of influences, then we would find that two influences acting together would produce a change equal to the product of the changes they would each produce in isolation. Note also that in this case the zero influence would produce a change of magnitude 1, which shows that magnitude in the space of influences is not necessarily the same thing as magnitude in the space of outcomes.

Here, I believe, we have hit upon the greatest limitation of the reductive method. The satisfaction of Assumption 5 guarantees the existence of an algebraic structure on the space of effects that can be used to calculate the effect that occurs when multiple influences are acting, but there is no guarantee that we will know what this algebraic structure is. Assumptions 4 and 5 place some restrictions on the possible algebraic structure, and we can further restrict the possibilities by, say, empirically learning what the 'zero effect' is and learning what the result of combining some elements is. Typically, however, evidence of this sort will still leave the algebra underdetermined—There will be many possible algebras consistent with the evidence. In practice, if we wish to extend our reductive explanation to cover a previously unstudied case, we must hypothesize that the algebraic structure of composition can be represented by some well-known mathematical structure, such as the algebra of vectors under vector addition.

5.6 Are Influences Redundant?

The reductive method, as described above, assumes that the observable behaviour of some system is determined by the resultant influences acting on the objects that make up the system. Furthermore, these resultant influences are determined by the basic influences that are manifested in the system, and the basic influences are determined by the state of the system at the time. In short, the state of the system at a time determines the influences at play, and these influences in turn determine the observable behaviour at that time.[7]

These assumptions of the reductive methodology imply that, given any system, there will be a function which maps states of that system directly to observable outcomes (or probability distributions over outcomes). Call this the 'total behaviour function' for the system. If we know this function, then we will be able to predict and explain the behaviour of the complex system without making any reference to influences. But if this is so, then one might argue that influences are superfluous—they play no essential role, and so, in the name of simplicity they should be dumped. An ontology of influences, it would seem, implies its own redundancy.

In response to this objection, I concede that in any given case where it is possible to provide a reductive explanation as described above, it would also be possible to provide an 'explanation' that makes use of the total behaviour function, and does not mention causal influences. However, reductive explanations that do make use of causal influences have two great advantages over a 'total-behaviour-function' explanation. First, reductive explanations are modular: once we understand the basic influences manifested within a simple system, we can use this knowledge to help explain what is going on whenever systems of that type occur—even if they occur within a more complex system. Thus, the reductive method gives us a way to tackle very complex systems and systems that are composed from known parts in novel arrangements. The total behaviour function, on the other hand, is determined by the system as a whole and—if we do not apply the reductive method—this function can only be learned by studying the whole system. What is more, this knowledge can only be applied in cases where the whole system is of exactly the same type. Each time we were faced with a different arrangement of parts, we would have to start from scratch to figure out the total behaviour function.

Second, not only are reductive explanations more flexible than total-behaviour-function explanations: they are more 'explanatory'. The modular nature of reductive explanations means that many different systems can be explained in terms of the same basic causal powers to manifest basic influences. In this way reductive explanations tend to unify our understanding of diverse phenomena. This kind of unification is something that Friedman (1974) and Kitcher (1989) have argued is constitutive of explanation. What is more, a number of philosophers (e.g. Salmon (1984)) have argued that causation is an essential part of explanation, and—as I will argue in Chapter 8—causal influences are causal in more than name, so reductive explanations are indeed causal explanations.

[7] In the probabilistic case, one or both of these steps involves chance: the state of the system might determine only a probability distribution over possible influences, and the influences at play might determine only a probability distribution over the possible behaviours of the system.

5.7 Conclusion

In this chapter I have attempted to give a detailed and reasonably formal account of the reductive method, with the aim of identifying a number of assumptions of the method. For convenience, I present these assumptions again below:

Supervenience When attempting a reductive explanation of some phenomenon P in complex system S, we assume that there is a set of basic objects of S such that the relevant properties of P supervene on the properties of, and relations between, these basic objects.

Limited Rank When attempting a reductive explanation of rank N of some phenomenon P in complex system S, we assume that there is an explanatory basis for P in S that does not include any basic powers that have a rank greater than N.

Changes Determined by Influence The changes in any property P of any basic object a, at any time t, are determined by the set of all and only the basic influences that are individually directed towards changes of P in a at t.

Algebra of Composition Let I_{Pa} be the set of influences of type P acting on object a. There exists a two-place operator $\oplus$ such that:

1. The result of influences $u_1, u_2, \ldots$ and $u_n \in I_{Pa}$ acting together at the same time, and in the absence of any other influences in I_{Pa}, is the outcome that would occur if the influence $(\ldots (u_1 \oplus u_2) \oplus \ldots) \oplus u_n$ acted on its own;

2. $(I_{Pa}, \oplus)$ is a commutative semigroup.

Influence Characterized by Effect Causal influences are uniquely characterized by the effect that they would bring about if acting in isolation.

The first two of these assumptions concern the existence of an appropriate explanatory basis. Assumptions 3–5, on the other hand, are concerned with the nature of causal influences and the way that they combine to produce an outcome.

Looking at these assumptions, one could be forgiven for thinking that I have left something out. For in Chapters 3 and 4, I ascribed great importance to the requirement that the influences manifested by a causal power be invariant, in the sense that the power will always manifest the same influence given the state of the relevant subsystem; yet this invariance does not seem to be mentioned in the list. In fact, however, the invariance is built into the assumption of Limited Rank. If a causal power is of rank n, then the influence manifested by the power is invariant once we fix the state of an associated subsystem of n objects. So, if there is an upper limit to the rank of the causal powers involved in a reductive explanation, then these powers must be invariant given the state of their respective finite subsystems.

These assumptions are necessary for a perfect reductive explanation, but, as I suggested in Chapter 4, we can allow for less than ideal explanations in which one or more of the assumptions above are violated in some small way that makes no practical difference for the purpose of the explanation. If the assumptions are violated in a significant manner, however, then a reductive explanation of the phenomenon in question will not be possible. Nothing in the discussion so far allows us to rule out the possibility that these assumptions may fail to be satisfied, and I will consider the implications of such a failure of the reductive method in Chapter 10. Nonetheless, the great success and utility of the reductive method suggests that in many cases these assumptions are indeed satisfied and that we are justified in positing causal powers and influences of the type described here. So, in the remaining chapters, I wish to further investigate the work that an ontology of causal power and influence can do for us.

6

Macroscopic Power and Influence

In the eighteenth and nineteenth centuries, Canadian fur traders noticed an interesting phenomenon: every ten years or so there would be a significant drop in the population of snowshoe hares, followed shortly thereafter by a similar drop in the population of Canadian lynx (Krebs et al., 2001). The Hudson Bay Company, established in 1671, kept meticulous records of furs traded from across Canada, and analysis of this data shows a remarkably regular ten-year boom and bust cycle (Elton and Nicholson, 1942). The oscillations are so striking that they are featured in nearly all ecology textbooks (see, for example Townsend et al., 2008; Krebs, 2009). It is thought that the regularity of this cycle is due to the relatively simple ecology of the boreal forest: the snowshoe hare is the main diet of the Canadian lynx, and the lynx is one of the main predators of the hare, so that the cycles are due to a relatively pure predator–prey interaction. When the population of lynx is small, there is little predation upon the hares and their population increases. But as the population of hares increases, there is more food for the lynx and so the lynx population increases. Eventually, the large population of lynx starts to drive the population of hares down, but when this happens, there is less food for the lynx and so their population also decreases. Then the cycle starts again. Of course, there are other influences on the population of hares and lynx. Studies suggest that in order to explain the dynamics of the hare population we need to consider the influence of food supplies as well as the influence of predation by a number of species. On the other hand, the dynamics of the lynx population is due almost exclusively to the influence of the hare population (Stenseth et al., 1997; Krebs et al., 2001).

The explanation for the cycle of lynx and hare populations that has been sketched here is a reductive explanation. In this explanation, a property of the ecosystem (the hare population size, say) is explained by combining the influences exerted within various subsystems (the hare/lynx subsystem, the hare/vegetation subsystem, and the hare/other predator subsystem). And, as required by reductive explanation, these subsystems are treated as systems in isolation. Almost all ecology textbooks, for example, tell us that we can understand the 'underlying tendency' for cyclical dynamics by considering the predator–prey interaction in isolation. These texts all go on to discuss the Lotka-Voltera equations which provide a mathematical model of the predator–prey interaction in isolation from all other influences (see, for example, Townsend et al., 2008, 233–8; Krebs, 2009, 190–4; and Matthiopoulos, 2011, 243–4).

A central feature of the reductive explanation of hare and lynx cycles is that the basic objects involved are themselves complex entities. In particular, the behaviour being explained is a property (size) of a whole population of a species in a region, and the explanation involves interactions between whole populations of different species in that region. Clearly, however, the interactions between these populations are constituted by interactions between individual members of these populations. The whole population of lynx exerts a downwards influence on the size of the hare population because individual lynx eat individual hares. Thus, the powers and influences referred to in the reductive explanation of hare and lynx population cycles seem to be composite powers and influences–the powers of the lynx population are composed of the powers of the individual lynx, and the influence that the lynx population exerts is composed out of the influences exerted by the individual lynx.

Composite powers and composite influences are a common feature of reductive explanation. Indeed, outside fundamental physics it is likely that many, if not all, the powers and influences mentioned will be composite rather than fundamental.[1] In this chapter, then, I will explore the relation between composite powers and their components. In particular, I will consider how powers and influences can be composed both synchronically and asynchronically, and consider the implications of such composition with respect to avoiding antidotes and finks.

A major part of this chapter involves defending the idea that composite influences can be composed of more basic component influences. In particular, I defend the idea from objections that have been advanced by Andreas Hüttemann (2004), Jessica Wilson (2009), and Olivier Massin (2017) against the existence of composite and component forces. Since forces are paradigmatic examples of causal influence, I take it that these arguments will also be objections to my concept of composite influences more generally. Finally, having defended an account of the relation between composite and component powers, I will show that this account defuses an argument recently advanced by Alexander Bird (2016), who concludes that there are few, if any, true macroscopic powers.

6.1 Synchronic Composition of Powers

There are actually two very different ways that fundamental causal powers could come together to make up composite powers. First, there is what we might call

---

[1] Note that it is possible, in principle, for a composite entity to have a non-composite power. When viewed at the level of the component parts, such a power would appear as a high-rank power. See Chapter 10 for details.

synchronic composition, in which the component powers manifest at the same time to make a composite power. Second, there is asynchronic composition, in which component powers manifest one after another in a chain. I deal with synchronic composition here, and discuss asynchronic composition below.

Causal powers, as I have defined them, are dispositions to manifest a causal influence in response to an appropriate stimulus. Now, it is part of the concept of causal influence that, when influences of the same type act on the same object at the same time, they 'add together' to produce an outcome. I suggest, therefore, that we can understand synchronic composition of powers in terms of the composition of influences. This strategy for making sense of composite causal powers relies on the assumption that there can be composite causal influences, and I will defend this assumption in the next section.

To get an idea of how the strategy for composing powers works, let us start with a simple example of a single particle with two powers. Suppose, for example, that a is an electron; then a will have both a fundamental gravitational power to exert forces on other massive objects and a fundamental electromagnetic power to exert forces on other charged objects. Since forces are a single type of influence, they can be composed to form a resultant influence when they act on the same object at the same time. Thus, we could, if we so wished, view a as having a composite power to manifest this resultant force. More specifically, suppose that a is at a distance d from another fundamental object b, which has mass m and charge q. Then a will exert on b a gravitational force of magnitude $Gm_e m/d^2$ and an electromagnetic force of magnitude $\kappa_e q/d^2$, where G is the gravitational constant, m_e is the mass of an electron, κ_e is Coulomb's constant, and the forces are directed along a line from b to a. Since these two forces act on the same object at the same time, they will add according to the rules of vector composition to produce a resultant force of magnitude $(Gm_e m + \kappa_e q)/d^2$ directed towards a. Thus, we can treat a as having the composite causal power to manifest a force of magnitude $(Gm_e m + \kappa_e q)/d^2$ on any object of mass m and charge q located at a distance d from a.

In general, whenever an object a has two causal powers of the same type that each exert an influence on some object b, we can view a as having a composite causal power to exert the resultant influence on b. Since powers are represented by influence-valued functions, in order to prove that a composite power exists in this case we need to show that there is a function whose domain is the state-space of some system containing both a and b, and which outputs the appropriate resultant influence. There is nothing too conceptually difficult about this proof. However, it is a little technical, so I have separated it off below.

Proof. Suppose that a is a basic object with two causal powers P_1 and P_2 that are of the same type. Let S_1 and S_2 be systems of the type that are associated with P_1 and P_2 respectively, and suppose that a is an element of both S_1 and S_2, and that a is playing the role of the possessor of the respective power in each system.

This means that P_1 is represented by a function I_1 whose domain is the state-space of S_1, and that P_2 is represented by a function I_2 whose domain is the state-space of S_2. In particular, P_1 is the power to exert an influence $I_1(s_1)$ on some object $x \in S_1$, where s_1 is an element of the state-space of S_1. Similarly, P_2 is the power to exert an influence $I_2(s_2)$ on some object $y \in S_2$, where s_2 is an element of the state-space of S_2. Since these powers manifest influences of the same type, they would combine to form a resultant influence if they were to act on the same object at the same time. In particular, if b is an element of S_1 and of S_2, and if S_1 is in state s_1 and S_2 is in state s_2, and if a exerts two influences, $I_1(s_1)$ and $I_2(s_2)$, on b, then b will feel the resultant influence $I_1(s_1) \oplus I_2(s_2)$, where $\oplus$ is the appropriate influence composition operator, as described in Chapter 5.

We are almost in a position to define the composite power as a power to manifest this resultant influence. Note, however, that the two functions I_1 and I_2 do not necessarily have the same domain. In order to define the function that characterizes the composite influence, therefore, we must specify a domain for the function that includes the domains of I_1 and I_2. Let us define $S_1 \cup S_2$ to be the smallest system that contains both S_1 and S_2 (so it contains all, and only, the basic objects and properties of both S_1 and S_2). Since S_1 and S_2 are subsystems of $S_1 \cup S_2$, fixing the state of $S_1 \cup S_2$ will also fix the state of S_1 and S_2. So every state s of $S_1 \cup S_2$ will determine a unique state, $s|S_1$, of the subsystem S_1 and a unique state, $s|S_2$, of the subsystem of type S_2. Finally, note that any states s_1 and s_2, as defined above, must be compatible, since S_1 is in state s_1 at the same time that S_2 is in state s_2. Thus, given any such s_1 and s_2, there is a state s of $S_1 \cup S_2$ such that $s_1 = s|S_1$, and $s_2 = s|S_2$. Thus, we can represent the composite power of a to influence b with a function I whose domain is the state-space of $S_1 \cup S_2$, such that for any state s of $S_1 \cup S_2$, the composite power will be the power to manifest the influence $I(s) = I_1(s|S_1) \oplus I_2(s|S_2)$ on b when a and b are appropriate objects of the system $S_1 \cup S_2$. □

I have shown, then, that it is possible for a single basic object to have a composite power. Most macroscopic powers, however, will involve *composite* objects, with the component objects each contributing a component power. The gravitational power of the Earth, for example, is a composite power made up of the powers contributed by the many particles from which the Earth is constituted. In such cases we can indeed define a composite power and the strategy is reasonably straightforward: if the many component objects have powers of the same type, then the influences manifested by these powers will compose to form resultant influences, and we can view the composite object as having a composite power to manifest this resultant influence. The proof that this is possible is very similar to the proof above.

Proof. Suppose that x_1 and x_2 are objects such that x_1 has causal power P_1 and x_2 has causal power P_2 and suppose that these powers are of the same type. Suppose further that each of these powers exerts an influence on some object b. As above, let S_1 and S_2 be systems of the type that are associated with P_1 and

P_2 respectively, so that when S_1 is in state s_1, x_1 exerts an influence $I_1(s_1)$ on b and x_2 exerts an influence $I_2(s_2)$ on b. Then, when S_1 is in state s_1 and S_2 is in state s_2, b will feel the resultant influence $I_1(s_1) \oplus I_2(s_2)$. As was the case with a single object discussed above, the main difficulty in specifying the function that characterizes the composite influence is in stating the domain of the function. By almost identical reasoning to that given above, we can show that the domain for the composite power will be the state-space of $S_1 \cup S_2$ and the composite power will be the power to manifest the influence $I_1(s|S_1) \oplus I_2(s|S_2)$ on b when x_1, x_2, and b are appropriate objects of $S_1 \cup S_2$ in state s. □

Note that there is an important difference between a composite power of two objects and a composite power of just one. When dealing with a single object above, $S_1 \cup S_2$ had to include, as a minimum, a and b. If a and b are the same object, then the composite power, like its components, will be of rank 1. If a and b are distinct objects, then the composite power, like its components, could be of rank 2. In the case of a composite power of two objects, however, $S_1 \cup S_2$ must include at least x_1, x_2, and b. If b is the same object as either x_1 or x_2, the composite power will be of rank 2. If x_1, x_2, and b are all distinct, then the composite power is at least of rank 3. In general, composite powers of two objects will have a higher rank than the component powers from which they are composed. The practical implication of this is that it may not be particularly useful to keep track of such composite powers. For if the relevant properties of x_1 and x_2 can significantly vary independently of each other, then we must keep track of these properties individually in order to explain the behaviour of the system, and nothing is gained by treating the resultant influence as the manifestation of a composite power.

Nonetheless, there are a number of situations in which one or more of the relations between x_1 and x_2 will not change significantly. For example, if x_1 and x_2 are massive objects that have been glued together, then their distance from some third object b cannot vary independently, and it may indeed be useful to treat x_1 and x_2 as constituting a single object with a single distance from b, and a single composite gravitational power to influence b. Alternatively, if, during the period being investigated, the distance between x_1 and x_2 varies by only an insignificant amount compared to the distance between this pair and the third object b, then it could be useful to ignore this variation and again treat x_1 and x_2 as constituting a single object with a composite power to influence b. In general, if, for one reason or another, we can ignore one or more of the internal relations between x_1 and x_2, then this will reduce the size of the state space of $S_1 \cup S_2$. In such cases it may make sense to treat the two objects as a composite object with a composite causal power, since the reduced state space may be smaller than the combination of the individual state spaces of S_1 and S_2.[2]

---

[2] I have a hunch that this kind of situation is *constitutive* of our concept of a composite object: we treat a collection of individuals as a single object just when we are able to treat that object as having stable powers that are of lower rank than the collection of component powers taken individually.

The proofs above show that two powers (whether possessed by the same object or two separate objects) can combine to form a composite power. But this implies that three or more powers can also form composites. To see this, let us use the symbol + to signify composition of powers, so that $P = P_1 + P_2$ just in case P is composed from P_1 and P_2. Now suppose that three powers, P_1, P_2, and P_3, all simultaneously exert an influence of the same type on some object b. By the argument above, any two of them will form a composite power. So, in particular, $P_1 + P_2$ will be a composite power. But now we can use the argument above again to show that this composite power can itself compose with P_3 to form a composite power. So $P = (P_1 + P_2) + P_3$ will also be a composite power. Note that the commutativity and associativity of influence composition will transfer across to power composition, so $P_1 + P_2 = P_2 + P_1$ and $(P_1 + P_2) + P_3 = (P_1 + P_3) + P_2 = (P_2 + P_3) + P_1$. As we should expect, then, the identity of a complex power does not depend on the order in which its component powers are composed.

6.2 Composite Influences

Synchronic composition of powers is parasitic upon composition of influences. In particular, the proofs to the conclusion that $P_1 + P_2$ is itself a power rely on the assumption that $I_1 \oplus I_2$ is a causal influence. If it were not, then the functions defining $P_1 + P_2$ would not be influence-valued, and so $P_1 + P_2$ would not represent a causal power. Note that in Chapter 5, I defined $\oplus$ as an operator on the space of influences, and so we might respond that it is true simply by definition that $I_1 \oplus I_2$ is a causal influence. But this response avoids the real issue here. The claim that I wish to make is that when I_1 and I_2 act at the same time on the same object, they jointly constitute an influence that can be represented by $I_1 \oplus I_2$. This claim was simply assumed in Chapter 5, but it is not obviously true.

Consider, for example, a situation in which two people are shovelling sand into a wheelbarrow; the fact that the two people are acting together on the same object certainly does not imply that they jointly constitute a composite person. This is true even if the work they do is equivalent to the work that would be done by a single person shovelling twice as fast. Similarly, then, the fact that two influences act together on the same property of the same object does not necessarily imply that these two influences jointly constitute a composite influence. What is more, there are a number of recent arguments in the literature which suggest that two or more influences cannot jointly constitute a composite influence. The arguments I refer to deal with Newtonian forces rather than causal influences as I conceive of them. However, since Newtonian forces are paradigm examples of causal influences, any problem for forces is likely to also be a problem for causal influences more generally. Thus, it is important to address these challenges, and this will be the task of this section. Along the way, I will develop a positive account of why we should think that influences acting together jointly constitute composite influences.

6.2.1 The Overdetermination Argument

Jessica Wilson—developing an argument suggested by Creary (1981)—argues that if resultant and component forces are real and wholly distinct, this would imply a problematic causal overdetermination. Newton's second law tells us that the resultant force is sufficient to bring about the resulting motion. But the component forces, when taken together, are also sufficient to bring about the result (this is what grounds the appropriateness of explaining the resultant motion via vector addition of the component forces). Thus, accepting the reality of both resultant and component forces would lead us to conclude that every change in motion has two distinct, yet completely sufficient, causes. This conclusion, says Wilson, is unsatisfactory: 'Surely the effect in a given case of conjoined circumstances is not caused twice over—once by the component forces assumed to be present, and again by the (different) resultant force assumed to be present' (2009, 539–40).

Andreas Hüttemann (2004, 105) makes a similar point in a different way. Newton's second law, he says, tells us that the acceleration of a body is determined by the sum of all the wholly distinct forces acting upon it. Thus, if resultant and component forces are real and wholly distinct, then the acceleration of a body must be determined by the sum of the component forces *and* their resultant. But this conclusion is clearly absurd. This conclusion would effectively have us count the forces twice over, predicting an acceleration twice as big as that which actually occurs. Indeed, this conclusion may lead to an infinite regress, for, by the reasoning above, the vector sum of the component and resultant forces will itself be a distinct force, and so should also be included in the sum used to calculate the resulting acceleration. And so on. The conclusion that Wilson and Hüttemann draw is that we cannot attribute reality both to component forces and to resultant forces.

Wilson and Hüttemann's arguments show that we are led to absurdity if we assume that resultant and component forces are real and wholly distinct. But this leaves us with two options: we can follow Wilson and Hüttemann and deny that resultant and component forces are all real, or we can deny that resultant and component forces are wholly distinct (the third logical possibility—denying that component and resultants are all real *and* denying that they are wholly distinct—does not make much sense in this case). The second option, I suggest, is well motivated, and it will solve the problem of overdetermination.

Consider, by analogy, a car that is put together out of many components. We happily accept the existence of these components as well as the existence of the car that results when they are assembled. But the coexistence of the components and the resultant car does not lead to causal overdetermination. If someone is hit by a car, their injuries are not caused twice over, once by the car, and once by the components of the car. The reason that the coexistence of the car and its components does not lead to trouble is that the car is not wholly distinct from it components, for the components are related to the resultant as part to whole.

We are happy to admit the reality of both parts and whole because, in some relevant sense, the whole is nothing over and above the sum of the parts. Since the whole is constituted by the parts, the existence of the whole simply is the existence of the parts arranged appropriately. Similarly, a piece of chocolate has numerous ingredients, and a symphony is composed of a number of movements. In each of these cases, along with innumerable others, we readily admit the existence both of the components and of the entity that results from their composition.[3]

But Nancy Cartwright has argued that the composition of forces does not fit this model. Component forces, she says, are not related to resultant forces as parts to whole. Cartwright's argument for this proceeds as a response to John Stuart Mill. Mill suggests that component forces always have their full effect, with the individual accelerations produced by these forces being literal parts of the overall acceleration:

> If a body is propelled in two directions by two forces, one tending to drive it to the north and the other to the east, it is caused to move in a given time exactly as far in both directions as the two forces would separately have carried it; and is left precisely where it would have arrived if it had been acted upon first by one of the two forces, and afterward by the other.   (Mill, 1843, III.VI.1)

Cartwright objects as follows:

> Mill's claim is unlikely. . . . When a body has moved along a path due north-east, it has travelled neither due north nor due east. The first half of the motion can be a part of the total motion; but no pure north motion can be a part of a motion which always heads north-east. . . . The lesson is even clearer if the example is changed a little: a body is pulled equally in opposite directions. It doesn't budge an inch, but on Mill's picture it has been caused to move both several feet to the left and several feet to the right. . . . It is implausible to take the force due to gravity and the force due to electricity literally as parts of the actually occurring force.
>
> (Cartwright, 1980, 79)

Cartwright is surely correct that a lack of motion cannot be literally composed of a motion to the left simultaneous with a motion to the right. But this is an argument against the reality of component motions, not against the reality of component forces.

---

[3] A small minority of philosophers do, in fact, deny the existence of composite entities like cars and chocolate bars (see, for example, Peter van Inwagen 1990 and Trenton Merricks 2001). I do not have space to consider their arguments here, but I note that Merricks's main argument against composite objects is an overdetermination argument much like Wilson and Hütterman's. Merricks argues that if composite objects existed, they would either be epiphenomenal, with no causal powers at all, or any effects they have would be caused twice over: once by the composite object and once by its constituent parts. Both possibilities, he says, are objectionable. If I succeed in defending the reality of both composite and component influences below, then I think this defence might also be used to launch an objection to Merricks's claim that any causation by composite objects must be objectionably overdetermined.

But perhaps we can view Cartwright's argument as an argument from analogy; just as an acceleration with a particular magnitude and direction cannot be literally made up of parts that have different magnitudes and directions, a force of a particular magnitude and direction cannot be composed of parts with different magnitudes and directions; just as a lack of acceleration cannot have an acceleration to the left and an acceleration to the right as parts, a lack of force cannot have a force to the left and a force to the right as parts. The analogy turns on the fact that component accelerations and component forces are vectors, and unlike scalar quantities, vectors can add to produce resultants that have a direction different from any of the components. Perhaps, then, we should be reading Cartwright's discussion of component accelerations as a particularly vivid example of how vector addition differs from scalar addition.

But there is an important disanalogy between accelerations and influences such as forces: it is part of the very concept of an influence that they combine somehow to produce an outcome, but the same is not true of accelerations. Accelerations do not *produce* outcomes: they *are* outcomes; and this is why vector composition in the case of accelerations does not seem truly to be addition. Suppose that we place two identical positively charged massive spheres near to each other in an otherwise isolated environment. Suppose also that the charges of the spheres are such that they end up accelerating towards each other at exactly half the rate they would have if they had not been charged. In this situation it seems perfectly natural to say that sphere *A* exerts *two* forces on sphere *B*: one gravitational and one electromagnetic. But it is also perfectly natural to speak of *the* force exerted by *A* on *B*. We might, for example say that the force *A* exerts on *B* is exactly half what it would have been if the spheres were not charged. How can we reconcile these apparently contradictory, yet natural claims? How can the number of forces that *A* exerts on *B* be equal to two and equal to one? Cartwright and Wilson would have us deny one of the natural claims (the first), but we can accommodate both claims if we simply assert that the singular, overall force that *A* exerts on *B* is somehow constituted by the gravitational and electromagnetic forces (as wholes are constituted by their parts, for example). The apparently conflicting claims about the number of forces would then be no more mysterious than the apparently conflicting claims that a car is a single object but is also a collection of many objects. Unlike the case of accelerations, then, there does not seem to be anything absurd in the claim that resultant forces are somehow constituted by their components. Indeed, this claim seems to be motivated by our natural intuitive ways of describing forces. Thus, the analogy between accelerations and forces breaks down and Cartwright's argument against the possibility that component forces are related to resultant forces as parts to whole fails.

Wilson (2009, 541) shares Cartwright's belief that it makes no sense to think of component forces as parts of resultant forces, but she provides a slightly different argument. Our ordinary understanding of parthood, she says, at least in so far as it is applied to 'broadly scientific entities', involves dividing these entities spatially or temporally. But component forces are neither spatial nor temporal parts of resultant forces. For example, if the gravitational and electromagnetic forces exerted by sphere A on sphere B in the example above have any spatial or temporal location at all, then surely they must have the same spatial and temporal location, since they are exerted at the same time by the same object on the same object. Thus, concludes Wilson, the relation between component and resultant forces cannot be anything like the ordinary part–whole relation. Nor, she says, is there any way to extrapolate the ordinary part–whole relation to the context of forces. We might be able to perform such an extrapolation in the particular case where a component force F_c has a magnitude less than the resultant force F_r and both forces are exerted along the same direction. But there is no way to extrapolate the part–whole relation to a situation in which component forces have a direction different from the resultant, or have a magnitude greater than the resultant. Wilson concludes by agreeing with Bertrand Russell that '[vector] composition is not truly addition, for the components are not parts of the resultant. The resultant is a new term, as simple as its components, and not by any means their sum' (Russell, 1903, 477).

There are actually two separate obstacles identified by Wilson in this argument. The first is that the magnitude of a force—or any influence—is an *intensive*, as opposed to *extensive*, quantity. Extensive quantities, in the sense that I use the term here, are those like mass or volume, whose total value is the sum of contributions made by spatial parts. If you take a 1 kg mass, for example, and divide it in half, you end up with two half-kilogram masses. There is a clear sense in which we can say that the original 1 kg mass was literally composed of the two half-kilogram masses. The value of an intensive quantity, like density, temperature, or redness, on the other hand, does not depend on spatial parts in this way. Take an object that has a density of 1 kg/m^3 and cut it in half; you end up with two objects, each of which has a density of 1 kg/m^3—not $\frac{1}{2}$ kg/m^3 (assuming an even density distribution). Thus, there does not seem to be a sense in which the density of the original object is the sum of the densities of the two halves.

In general, it does seem difficult to make sense of the idea that intensive quantities such as temperature or redness have parts. Two electric heaters may contribute to the temperature of a room, for example, but we cannot make sense of these two contributions existing simultaneously; the temperature of the room is not literally composed of two component temperatures. The problem is that there is no way to separate out the two components. We cannot think of them coexisting 'side by side', since there is no spatial or temporal separation of the two components, so if the two component temperatures really exist, then the room would have to have at least two different temperatures simultaneously, and this, surely, is impossible (we might worry that the room would, in fact, have to have *three* different temperature simultaneously: the two component temperatures and the resultant).

This same problem does not arise, however, in the case of forces. When we claim that two component forces really exist, we are not thereby attributing two logically incompatible properties to a single object, as is the case if we claim that component temperatures exist. The component forces are separate objects and can thus have separate properties. Of course, the component forces are not separated spatially, but they are separated within the vector space in which they exist. Separation within the vector space of forces is, admittedly, harder to visualize than separation in space, but it is separation nonetheless. Indeed, in his pioneering work in the development of the theory of vector spaces, Hermann Grassmann conceived of a vector space precisely as describing the structure of a generalized concept of extensional magnitude. Grassmann called his new theory *Ausdehnungslehre*, 'Extension Theory', and in his more philosophical 1844 publication on the subject he says:

> in order to obtain the extensive magnitude I relate it to the generation of the line. Here it is a generating point that assumes a continuous sequence of positions; and the collection of points into which the generating point is transformed with this evolution forms the line ... In the same way we now proceed to the extension in our science, if we simply replace these spatial relations by their corresponding concepts. (Grassmann, 1995, 46)

Grassmann's talk of generating a line through evolutions of a point captures the idea that the line is made up of point-like parts. He then develops extension theory as a study of the abstract structure of this part–whole relation.

Thus, the fact that forces are intensive in the spatial (and temporal) sense does not mean that they cannot be extensional in some more general sense, and hence does not mean that it is incoherent to speak of component forces being literal parts of resultant forces.

Grassmann's work also provides a response to Wilson's second—and perhaps main—problem with extrapolating the part–whole relation to cover forces. Her objection is that because forces are vectors, with both a magnitude and a direction, a resultant force can have a component that has a greater magnitude than the resultant, or that points in a completely different direction from the resultant. Wilson is correct that it is hard to see how such a component can be a part of the resultant, but, as mentioned above, Grassmann's development of the theory of vector spaces was precisely a generalization of the concept of extensional magnitude beyond the realm of the spatial. The vector-space structure which forces satisfy, therefore, just *is* an extrapolation of the part–whole relations we find in spatially extended objects.

Wilson is simply wrong, then, in her claim that there is no way to extrapolate from the part–whole relation as it applies to spatial components to a part–whole relation between vector components and resultants, for the entire vector calculus was first constructed through such an extrapolation. However, there may still be something to Wilson's argument since this extrapolation may take the part–whole relation so far from its original interpretation that it no longer provides a way to avoid the threat of overdetermination. As Wilson puts it:

> the point of appealing to the part/whole relation is to show how the threat [of overdetermination] may be avoided in non-eliminativist fashion: if the whole (the resultant force) is the sum of its parts (the component forces), then—by analogy to ordinary cases where wholes are reductive sums of parts, as in the case of lengths or masses—we can see how component and resultant forces can jointly exist without inducing overdetermination. But if the intuitive understanding of parthood does not make clear sense as applied to resultant and component forces, then appeal to this relation loses its dialectical force, being little better than a dogmatic insistence that no problematic overdetermination is at issue.
>
> (Wilson, 2009, 542)

So, is there any reason to believe that the part–whole relation, *as it applies to vectors*, will allow us to avoid the threat of overdetermination? The answer is yes, because, as we shall see, the relationship between resultant and component vectors is a case of the relation between determinable and determinate.

6.2.2 Resultants as Determinables

Recall that Assumption 5 of the reductive method states that influences are characterized by the outcome towards which they are directed (see Chapter 5). Forces, therefore, will be characterized by the accelerations towards which they are directed. In particular, to be a force of magnitude x and direction $\mathbf{u}$ *just is* to be an influence directed towards an acceleration of magnitude x/m and direction $\mathbf{u}$ when acting on an object of mass m. Now, the existence of an influence does not guarantee any particular outcome, since the outcome depends on all the relevant influences; however, when an influence acts in isolation (or at least when there are no other significant and relevant influences at play), the outcome to which it is directed is guaranteed. Putting these two ideas together we can come up with a functional definition of what it is to be a force:

Definition 15 (Forces). To be a force of magnitude x and direction $\mathbf{u}$ *just is* to bring about an acceleration of magnitude x/m and direction $\mathbf{u}$ when acting in isolation on an object of mass m.

Now, suppose we assume that component forces exist, and suppose that two such forces $\mathbf{F}_1$ and $\mathbf{F}_2$ are acting on some object of mass m. Let $\mathbf{F}_T$ denote the vector sum of $\mathbf{F}_1$ and $\mathbf{F}_2$, so $\mathbf{F}_T = \mathbf{F}_1 \oplus \mathbf{F}_2$. Newtonian theory tells us that if these are the only forces acting on the object, then the object will undergo an acceleration of magnitude $|\mathbf{F}_T|/m$ in the direction of $\mathbf{F}_T$. But according to our functional definition, this is what it is for there to be a force equal to F_T acting on the object. Thus F_1 and F_2 acting together instantiate the force F_T. In general, the joint existence of component forces acting on an object will instantiate the resultant force. Resultant forces, then, are related to component forces as functional type to realizer, and this is just a variety of the determinable/determinate relation.

One might object at this point that I am being inconsistent. For I argued above that components and resultants were related as parts to whole, and now I am saying that they are related as determinant to determinable; and surely they cannot be both. But, at least in the sense that I am concerned with here, they can indeed be related in both ways. To see this, consider an example from the realm of our ordinary, intuitive concept of spatial parts. Suppose that a bodybuilder wants to bench-press 70 kg, and suppose that the bar alone has a mass of 10 kg. Then, the bodybuilder needs to add 60 kg to the bar. Now there are many ways that she could do this; she could, for example, add two 20 kg weights and two 10 kg weights, or four 15 kg weights, or six 10 kg weights. However she does it, she will end up with a barbell that has a mass of 70 kg. Thus, *being a barbell with a mass of 70 kg* is a determinable property which can be instantiated in many different ways. But each instantiation is itself a collection of components that are related to the 70 kg barbell as parts to whole. Just as the collection of bar and weights appropriately arranged both *constitute* and *instantiate* a 70 kg barbell, component forces appropriately arranged (i.e. acting on the same object at the same time) both constitute and instantiate a resultant force.

Wilson (2009, 543) considers and rejects the idea that resultant forces are determinables of more determinate component forces. First, she points out that determination is a relation of increased specificity, while forces (whether resultant or component) are all equally specific. However, she does acknowledge the 'abstract' possibility that we could view resultant forces as being less specific than (sums of) component forces, since many different sets of component forces could add to the same resultant. This is precisely the possibility that I have argued for above. What is more, there is nothing particularly 'abstract' about this possibility, for the situation is not really any different from the situation involving good old, concrete, spatial components. There is a sense, for example, in which different masses are all equally specific. Nonetheless, as we saw above, when we consider complex objects, there is a sense in which the mass of the object is a determinable, since there are many different sets of component masses that could be added together to produce a given total mass. So long as we admit the reality of the components (whether they be masses or forces), there is nothing strange or unintuitive about viewing resultants as determinable in this way.

Wilson has two more objections to the idea that resultant forces are determinables. She argues that if resultants were indeed determinables, then the determinant instance could not be simply the set of component forces: it must be their sum. For the resultant force has a unique magnitude and direction, but a mere conjunction (as opposed to a sum) of component forces does not. This leads to a problem in the case of a single force acting in isolation. In such situations there is a resultant force, but there are no more specific component forces, and hence no sum of component forces to give a determinant value to the resultant. But then we would have an instance of a determinable which does not take on any determinant value, and this is impossible ('nothing can be red without being a specific shade of red' (2009, 543)).

Wilson is right that a resultant is not instantiated simply by a set of components, but she packs too much into the notion of 'summation' when she claims that a single force cannot constitute a sum. All that is required to instantiate a resultant force is a set of component forces appropriately related. In particular, the appropriate relation is just that they be acting on the same object at the same time. So there is nothing to stop the instantiating set from consisting of a single force. In such situations in which there is only a single force acting, the resultant force is simply identical to the single component that instantiates it.

Wilson's final complaint is that even if we call the relation between resultant forces and components 'determination', this will not guarantee that we avoid the threat of overdetermination. In order to avoid this threat, she says, the relation between resultant and component forces must satisfy the condition that every causal power of the resultant is numerically identical with a power of the component forces when they are jointly present. But, she says:

> to suppose that resultant forces are determinables (or whatever) of sums of component forces does not guarantee satisfaction of the condition, since the status of the summation relation remains crucially unclear, insofar as component forces are not appropriately taken to be existing parts of resultant forces.
>
> (Wilson, 2009, 543–4)

There are three problems with this argument. First, it does not make sense to insist that the causal powers of the resultant force are numerically identical to the causal powers of the components taken together. For forces are influences, and hence are manifestations of causal powers. They do not themselves have causal powers. The only way we can make sense of Wilson's requirement, given an ontology of power and influence, is to argue that the outcome towards which the resultant is directed is numerically identical to the outcome towards which the components are directed when taken together. This requirement is indeed satisfied.

Second, the summation relation is not as unclear as Wilson makes out. As I pointed out above, the summation of component vectors is indeed an extrapolation of the intuitively understood summation of spatial components.

Third, the determination relation between component and resultant forces is the relation between realizer and functional type, and this relation is quite familiar, and will clearly allow us to avoid the threat of overdetermination. It is true that the acceleration of an object is brought about by the resultant force acting on that object. It is also true that the acceleration of the object is brought about by the joint action of the component forces acting on it. But the acceleration is not brought about twice over because the resultant force is realized by the components, and so is not distinct from the components. This situation with forces is no different from many others with which we are familiar. On 4 June 1940, the British prime minister gave a stirring speech to the House of Commons, assuring the world that Britain would stand firm against the threat of Nazi invasion. It is also true that this speech was given on 4 June 1940 by Winston Churchill. But the speech was not given twice over, because, at the time, the functional role of prime minister was realized by Winston Churchill and so prime minister and Churchill were not distinct. The causal powers of the prime minister are numerically identical to the causal powers of Churchill because the prime minister is numerically identical to Churchill. Similarly, the causal powers of the resultant force (if we wish to speak this way) are numerically identical to the causal powers of the component forces because the resultant is numerically identical to the components.

Having dealt with Wilson's overdetermination argument, I would now like to consider two arguments put forward by Olivier Massin (2017) against the view that I have sketched above. Massin actually agrees that we should grant reality to both resultant forces and component forces, and agrees that we can avoid the overdetermination argument by doing so. However, he argues that we cannot identify resultants with sums (or appropriately arranged sets) of components, but should instead identify resultants with the residues left over after opposing components cancel. I do not believe that Massin's arguments for his positive proposal stand up, though I will not consider them in any detail here. I will note, though, that since his picture allows the existence of both component and resultant forces, it would give us what we need to make sense of the reductive method. Thus Massin's positive view is an alternative way to avoid the objections of Wilson and Hüttemann.

6.2.3   Null Forces

Consider a situation in which two equal but opposite forces are acting on an object: should we say that there is no resultant force acting on the object, or should we say that there is a resultant force which is null (i.e. of zero magnitude)? If resultant forces are simply identified with sums of components (or, as I would have it, sets of components appropriately arranged), then it seems we are committed to the latter view, for the sum surely exists in this situation just as much as it would if the two forces were not equal. As Massin puts it, 'a non-existent cannot be composed of existents, be it vectorially or otherwise' (2017, §4.3). So, the view I am endorsing is committed to the existence of null forces. But, Massin argues, we have good reason to deny the existence of such forces.

Massin gives us two reasons for rejecting null forces. First, he suggests that null forces are somewhat counter-intuitive, and have even more counter-intuitive consequences. When considering an object of zero weight, he says, 'we are tempted to say that such things have no weight rather a weight of 0, in the same way that we tend to think of things of zero height as having no height' (2017, §4.3). Furthermore, if we accept the existence of null forces, then Newton's second law would seem to imply that we should also accept the existence of null accelerations, and these null accelerations will imply further null quantities such as null impulses (changes in momentum). And if we are happy to accept null resultant forces, he asks, why should we not also accept null component forces? Should we not, for example, say that two massless particles will exert a null gravitational force on each other? Physical reality, he says, 'ends up being filled with zero accelerations, zero masses, zero interactions . . . That certainly sounds odd' (2017, §4.3).

The second—and main—objection that Massin raises against the existence of null forces, however, is that the very idea of a zero-value vector is incoherent. Vectors have both a magnitude and a direction, but there is no sensible, non-arbitrary, way to assign a direction to a vector with zero-magnitude. We could, he suggests, insist that a vector of zero magnitude also has a direction of zero. But what could this mean? Directions are usually measured as an angle from the abscissa. Thus, to say that a vector has zero angle is just to say that it points in the direction of the abscissa. But the direction of the abscissa is chosen arbitrarily as part of the coordinate system, and it would be nonsensical to insist that all null vectors in fact point in this direction. In general, it would be arbitrary to assign any particular direction to a null vector, and so, it seems, the only sensible thing to say is that the null vector has no direction at all. But, concludes Massin, if the null vector has no direction at all, then it is not a vector quantity. In particular, then, null forces cannot be vector quantities, and since forces are by definition vector quantities, null forces cannot be forces. Thus, the concept of a null force is incoherent.

Let us deal with Massin's main objection first. One could attempt a purely formal response. Null vectors are a perfectly coherent part of the mathematical structure of the vector space; indeed, it is part of the definition of a vector space that it must have such a null element. Thus, if the null vector does not have a direction, then it is simply false that all vectors must have a direction. So we could simply assert that null forces, unlike other forces, have no direction.

However, I believe that there is a more satisfying response. Forces, I have argued, are a species of causal influence, and such influences are, indeed, essentially directed; it is part of their nature that they point towards a particular outcome. But once we think of forces in this way, the problem of how to assign a direction to the null force disappears. For the null force simply points in the direction of the outcome in which there is no acceleration.

Massin's mistake is that he follows the common practice of thinking of forces as directed spatially, so that, for example, we can talk of forces being directed north-east, or south-west. In particular, this approach identifies the direction of a force with the direction of the acceleration that it tends to produce, and the magnitude of the force is identified as proportional to the magnitude of the acceleration that it produces. However, I have defined influences to be directed towards particular outcomes, not spatial directions. In the case of forces, the relevant outcomes are accelerations, which themselves have both a magnitude and direction. Thus, a particular force will not be directed spatially (e.g. northwards), but will be directed towards changes in velocity (e.g. towards a northwards acceleration of 3 ms^{-2}).[4] A null force, then, is simply a force directed towards no change in velocity.

The possibility of null resultant influences is even clearer outside the case of forces. Biological systems, for example, must maintain homeostasis, a state in which numerous internal variables, such as body temperature, are kept within an optimum range despite variations in the environment. When the environment exerts an influence that would change one of these internal variables, mechanisms within the organism must exert an influence that counteracts this. Ideally, the result of these external and internal influences acting together is no change at all in the relevant internal conditions of the organism. That is, the resultant is a null influence.

With this solution to Massin's main objection in hand, it is now easy to see how we should answer his minor objections. First, it is simply false that null forces have no direction. Null forces are influences that are directed towards no change at all, and there is nothing particularly counter-intuitive about such influences. Second, must we accept the reality of zero accelerations, zero impulses, and the like? No. There is no reason that we must consider a null force to be directed towards a real acceleration of zero magnitude, as opposed to no acceleration—a null force is simply a force directed towards no change in velocity. What of null component forces? Massin is correct that once we accept the reality of null resultant forces, we have opened the door to the possibility of null components, but nothing forces us to walk through this door. If there is reason to deny the existence of null component forces, then we could simply assert that only resultant forces can be null. This position could be justified, for example, if we thought that there was a set of basic force types (e.g. electromagnetic, strong nuclear, and weak nuclear), and we thought that all basic interactions of these types must have some positive strength.

There is no real problem with null resultant forces, and so commitment to such entities is no problem for the view I am defending here.

---

[4] The reason that it is so common to mistake the direction of a force for the spatial direction of its associated acceleration is that, as noted in Chapter 5, there is an isomorphism between forces and accelerations.

6.2.4 Apportioning Responsibility

The final objection I wish to consider was also put forward by Massin. He argues that views like mine, which identify resultant forces with sets or sums of appropriately arranged components, fail to correctly apportion causal responsibility. Massin (2017, 820–3) has us imagine a situation in which Tybalt is intentionally pushing Romeo towards a cliff, while Juliet is trying to prevent Romeo falling by pulling him in the opposite direction. Suppose that these pushes and pulls are the only forces acting on Romeo, and suppose that Tybalt is stronger than Juliet, so that Romeo falls to his death. Intuitively, says Massin, Tybalt is morally responsible for Romeo's fall partly because he is causally responsible for it. Juliet, on the other hand, bears no part of the causal responsibility for Romeo's fall—the force she exerted did not contribute to Romeo's acceleration towards the edge of the cliff— and so she bears no part of the moral responsibility.

The point of this tragic thought experiment is to bring out the idea that causal responsibility in Newtonian mechanics is usually understood to be apportioned (see Sober, 1988). In cases where multiple forces are acting on a body (as in the death of Romeo), it seems to make sense to ask what their relative contributions were to the resulting acceleration. But how is this done? The situation is straightforward when we only consider co-directional forces (in the spatial sense). Suppose, for example, that, instead of trying to save Romeo, Juliet is also pushing Romeo towards the cliff. Suppose also that Tybalt exerts a force of 20 N while Juliet only exerts a force of 10 N. Since both these forces are directed towards the cliff (or, rather, directed towards accelerations in the direction of the cliff), the resultant will be a force of 30 N directed towards the cliff. In this case it seems clear that Tybalt is responsible for two-thirds of the resultant force while Juliet is responsible for only one-third.

Things are not so clear, however, when we consider forces that are not co-directional. Consider again the situation in which Juliet is trying to save Romeo, so that her force of 10 N is directed away from the cliff. The total force is then 10 N (= 20 N − 10 N) directed towards the cliff. How do we apportion causal responsibility between Tybalt and Juliet? If we follow the method that was used in the co-directional case, then we get the answer that Tybalt is responsible for 20/10 of the resultant, whilst Juliet is responsible for −10/10 of the resultant. But this conclusion makes no sense, for portions can neither be negative nor greater than one. But if we do not have a way to apportion causal responsibility in this case, then it seems that all we can say is that Tybalt and Juliet were together responsible for the actual force of 10 N, and so together responsible for Romeo's acceleration towards the cliff. But this would remove our grounds for denying that Juliet has any moral responsibility for Romeo's death.

Now, says Massin, if a resultant force is simply identified with the set, or sum, of components, then we are left with no resources to apportion causal responsibility in non-co-directional cases. In the case of Romeo's death, all we can say is that there were two forces acting on Romeo, and these forces jointly constituted a resultant which produced an acceleration towards the cliff edge. Of course, we do know what accelerations the component forces would bring about in isolation, so we might think (as does Sober, 1988) that we could use this information to apportion causal responsibility. To make this work, however, we would need a way of apportioning the share that each component acceleration contributes to the resultant acceleration. But this just moves the problem from apportioning shares between component forces to apportioning shares between component accelerations. As Massin says:

> To determine the share of causal responsibility of component forces by means of the accelerations each would bring about in isolation, one would need to determine the share of each component acceleration in the resultant actual acceleration. But the vectorial composition of accelerations is mute with respect to such ratios. Apportionment is a division, and we cannot divide vectors by vectors (except colinear ones).   (Massin, 2017, 822–3)

I believe that the correct response to the thought experiments above is to simply deny that Newtonian mechanics gives us any straightforward way of apportioning causal responsibility. Massin believes that a correct apportionment of causal responsibility will attribute to Juliet no responsibility for Romeo's acceleration towards the edge of the cliff. But it would be wrong to think that this apportionment could simply be read off the structure of the composition of forces, since whether or not this is the correct attribution depends on the pragmatic context in which we asked the question. Suppose, for example, that instead of pushing and pulling Romeo, Tybalt and Juliet were exerting forces on a cart as part of a competition in which teams of two had to get their cart to accelerate at a given rate. Tybalt exerts a force of 20 N to the left, while Juliet exerts a force of 10 N to the right. The resultant force of 10 N to the left is exactly what is required to achieve the necessary acceleration, and they win the competition. It would hardly be fair for Tybalt to claim all the prize money for himself on the grounds that the force Juliet exerted did not contribute to the cart's acceleration. The desired acceleration would not have been achieved if either one of them had not exerted the force they did. Nor would it seem fair for Tybalt to claim more than half of the prize money on the grounds that he was more causally responsible for the resultant acceleration than was Juliet. In this context, it seems correct to simply say that Tybalt and Juliet both contributed to the cart achieving the desired acceleration and there is no sensible way to apportion responsibility between them any more precisely.

In other contexts, however, the asymmetry between Tybalt's and Juliet's contributions may be relevant. Tybalt may, for example, argue that he needs to eat more than Juliet to refuel after the competition since he expended more energy than she did. In the original thought experiment the asymmetry between Tybalt and Juliet is important because we are not interested in their contributions to the particular acceleration suffered by Romeo, but instead we are interested in their contributions to the fact that this acceleration was towards the cliff edge. So, we might be concerned with questions such as 'Would Romeo have fallen off the cliff if Tybalt had not been pushing him?' or 'Would Romeo have fallen off the cliff if Juliet had not been pulling him?' Newtonian mechanics, given any interpretation of the metaphysics of vector addition, answers 'No' and 'Yes' respectively. These answers explain and justify the intuitive thought that Tybalt, but not Juliet, was responsible for Romeo's death.

The point here is that Newtonian mechanics can give the right answers about who is responsible for what without having to apportion causal responsibility in situations involving forces that are not co-directed. We just need to be careful about the question that is being asked.

6.3 Asynchronic Composition of Powers

So far, I have considered composition of influences in cases where two or more influences of the same type are acting on the same object at the same time. However, there is another quite different way in which influences may join together to produce a resultant influence, and hence another way in which we might get composite powers. In particular, multiple influences can behave like a single influence by forming a chain. Unlike the component influences involved in the discussion of composition above, the intermediary influences making up a chain need not be of the same type; they are typically acting on different objects, and they are not exerted at the same time.

Consider, for example, what happens when you press your foot on the brake pedal in a car. The force applied to the pedal is transferred to the master brake cylinder, where it is transferred to pressure in the brake fluid that feeds the brake lines. This pressure is then transferred to force in the slave cylinders, and this force is transferred to the brake calipers, which press the brake pads against the discs. This produces a friction force which decelerates the car. So, the brake pedal exerts a decelerating influence on the car when it is pressed, but this decelerating influence is constituted by a long chain of intermediary influences. We could, of course, analyse this chain even further. Each step in the chain is also constituted by a chain of more fundamental influences, bottoming out, presumably, in chains of electromagnetic influence and strong nuclear influence between the electrons, protons, and neutrons that make up the components of the car's braking system.

Other examples of chains of influence are easy to find: the influence of the population of lynx on the population of hares, the influence a bag of apples exerts on the pointer of a scale upon which it is placed, the influence of an antibiotic over the course of a disease, and the influence of interest rates on the rate of inflation all seem to be constituted by chains of more fundamental influences. Indeed, most influences at the macroscopic level are constituted by such chains. So, if reductive explanation is to work in the way I have described at the macroscopic level, then these chains of causal influence must themselves be able to play the role of causal influences in reductive explanation.

Consider again the brake pedal. Clearly, it will not have the power to exert a decelerating influence on the car when pressed unless a whole host of other conditions are satisfied: the pedal must be correctly installed, the brake lines must be intact and full of brake fluid, and so on. But, given that these conditions are satisfied, pressing on the pedal will (normally) produce a decelerating influence— I will discuss the need for the qualification 'normally' shortly. Thus, it is possible to give a functional description of the pedal's decelerating power, but a complete description will need to specify that all these other conditions of the car are satisfied. Since there are many objects involved in these conditions, the causal power will be of a fairly high rank, and may have fairly complex triggering conditions. Nonetheless, for the purposes of explanation, we could choose to ignore the fact that there is a chain involved, and simply treat the pedal as having the power to exert a decelerating influence when all of the appropriate conditions (including the depressing of the pedal) are satisfied. Or perhaps we should say it is the whole braking system that has the power to exert a decelerating influence when the pedal is pressed; after all, the triggering conditions for the power will refer to components of the braking system well downstream of the pedal.

6.4 Approximate Powers: Keeping Things Simple

Composite causal powers—whether they are constituted by component powers acting together or in chains—will typically involve a large number of variables. The influence manifested by each component power will depend on its own set of variables, and the set of variables required to determine the influence of the composite power will be the union of all these component sets. However, powers that involve a large number of variables can be rather impractical to employ in day-to-day explanations and predictions. What is more, the higher the number of objects and variables involved in specifying a causal power (that is, the larger the state-space associated with the power), the less useful the power will be in explaining what is going on in new situations, since the power can only be employed in situations that involve all the associated variables. In general, therefore, powers that involve fewer variables will be preferred.

Fortunately, there are many situations in which we can ignore some of the variables that would be involved in a full specification of a composite causal power. The full specification of the power of a brake pedal to exert a decelerating influence on a car, for example, would refer to all the variables involved in all the powers along the chain from pedal to brake pads. Nonetheless, when a car is running normally, it is almost always the case that pressing the brake pedal will manifest a decelerating influence on the car. This is because the car has been carefully designed and built so as to ensure that all the other conditions required for the manifestation of this influence are satisfied. As Cartwright (1999, 50) might put it, the car is a 'nomological machine', designed to create a stable and regular connection between pressing the pedal and a decelerating force. Because of this regularity, we can, in normal circumstances, treat the brake pedal as a basic object that has a basic power to exert a decelerating influence when it is pressed. We can then refer to this causal power when making predictions, or providing explanations of the car's behaviour.

The stability of the relationship between pedal pressing and decelerating forces allows us to simplify our description of the brake pedal's causal power. It also means that we can treat brake pedals in different vehicles, which may work in very different ways, as having the same power. So long as the nomological machine is running properly, we can treat a brake pedal as having the power to exert a decelerating influence regardless of whether this power is constituted by a chain of powers associated with a hydraulic braking system, an electronic braking system, or another system entirely.

In general, when specifying a causal power, P, in some context C, we can ignore any variables that either (a) do not vary significantly within context C or (b) do not make any significant difference to the influence that is manifested by P in C. Of course, under-specifying a power in this way will not produce a description of a true power as defined in Chapter 4, since the description will be open to counterexamples, especially if we step outside context C. Although a brake pedal may have the power to decelerate a car when depressed in normal contexts, it will not produce such a deceleration if the brake lines have been cut. So long as we stick within context C, however, the under-specified description of P will do for practical purposes. The power so described will be an approximate power, as defined in Chapter 4, satisfying SCAP*, and therefore useful in reductive explanation.

Treating the pedal as having this power is good enough in normal circumstances, but when the causal power fails to behave as expected, then we need to pay attention to the more fully specified power, and ask whether any of the other conditions have not been satisfied. Typically, figuring out all the variables involved in the full specification of a composite power (whether it is constituted by component powers acting together or in a chain) will involve paying attention to the more basic powers from which the composite power is composed. So, to figure out why the brakes on a car have stopped working, we need to know the specifics of the chain of powers involved.

In general, if a previously observed regularity starts to break down in certain circumstances, one possible explanation is that the regularity is due to an approximate power that is constituted by more basic powers. If we can figure out what these more basic powers are, then we can explain both why the approximate power holds in normal circumstances and why it breaks down in other circumstances. Another possibility, however, is that the fully specified causal power involves more variables than expected, but is, nonetheless, not itself composed of more basic powers (the 'high-level' or 'emergent' powers to be discussed in Chapter 10 would be examples of such complex basic powers). Since powers that involve fewer variables are more explanatorily useful, it is good methodology to try to explain complex powers that involve many variables as composite powers that are composed from simpler, more basic powers. It is this methodological fact, I believe, that drives reductionism. Of course, nothing I have said so far guarantees that the search for such reductive explanations will be successful. So far, we may have simply been lucky.

6.5 Antidotes, Finks, and Composite Powers

In Chapter 4, I argued that basic causal powers could be given a simple conditional analysis, and that the associated function, or conditional, would not be open to counterexamples. As we have just seen, if a power is not fully specified, so that it is described by a function or conditional that does not include all of the variables relevant to that power, then it will be possible to find counterexamples—all we need do is find a situation in which one of the relevant unmentioned variables does not take on one of its 'normal' values. Thus, powers that are only described approximately will give the appearance of being susceptible to finks and antidotes. I say 'give the appearance' of being susceptible, because any power of finite rank can, in principle at least, be fully specified by a function, or conditional. This fully specific description will not be open to counterexamples, and so the fully described causal power will not be open to finks or antidotes.

The conclusion that I have just drawn, that fully specified causal powers are not susceptible to finks and antidotes applies equally to basic powers, and to powers composed from other powers acting simultaneously. However, it does not strictly apply to powers composed from chains of component powers. To see the problem, imagine that the braking system of a car involves a chain of just two causal powers. The pedal has the power to exert a closing force on the brake calipers when pressed, and the brake calipers have the power to exert a decelerating force on the car when they are closed. Suppose, further, that these component powers are fully specified: If the brake pedal is pressed, it will, without fail, exert a closing force on the brake calipers, and if the calipers are closed, they will, without fail, exert a decelerating force on the car. It is still possible that there are situations in which the brake pedal is pressed but no decelerating force is produced. As discussed in Chapter 4, although an influence may be directed towards a characteristic outcome, the mere presence of that influence does not guarantee the outcome. So, although pressing the brake pedal is guaranteed to produce an influence that is directed towards the closing of the brake calipers, this influence does not guarantee that the calipers will actually close, and so pressing the pedal does not guarantee that the calipers will manifest a decelerating influence. Whether or not the calipers close will depend on all of the influences acting on the calipers at the time. If, for example, the hinges on the calipers have been welded into an open position, then the force produced by pushing on the brake pedal will be counteracted by electromagnetic forces within the weld with the result that the calipers will not move into a closed position.

In general, each causal power in a chain is only able to exert an influence on the next. But what the next power does depends on the state of its associated subsystem, and this state depends on *all* of the influences that act on it, not just on the influence produced by the previous power in the chain. It is, therefore, always possible to interfere with the operation of a chain of causal powers by adding an appropriate influence, as was done by welding the brake calipers. We could, perhaps, try to remedy this problem by including the absence of all interfering influences as part of the full specification of a composite causal power. But, as Hartry Field (2003) points out, to ensure the absence of all interfering influences we would need to specify the state of an entire cross section of the past light cone of the system. A power this specific will involve a huge (possibly infinite) number of variables and so will not be much use in reductive explanation. Thus, a macroscopic causal power that is constituted by a chain of causal powers can only ever be approximately described. Power-chains may be treated as causal powers for the purpose of explanation in normal circumstances, but they will always be open to finks and antidotes, and so they can only approximate the conditions required for ideal reductive explanation.

6.6 Bird on the Existence of Macro Powers

In a recent article, Alexander Bird (2016, 2) defines what he calls the macro powers thesis:

Macro Powers Thesis (MacroPT) Many or all macro properties are powers or clusters of powers and such properties play a role in explaining important phenomena involving macro entities.

Here 'macro' is simply shorthand for 'non-fundamental'. MacroPT, says Bird, is assumed by authors like Mumford and Anjum (2011a) who believe that an ontology of powers can help address philosophical problems regarding, for example, causation, representation, intentionality, free will, and liberty. Something like

MacroPT is also assumed by my claim that reductive explanation can occur at non-fundamental levels, since my account of such reductive explanation will involve non-fundamental causal powers.

Bird argues that powers are justified at the fundamental level by the role they play in making sense of cross-world identity of properties and in explaining the laws of nature (see Chapter 7 for more discussion of the latter). However, he argues that our justification for believing in fundamental powers does not automatically give us reason to believe there are powers at the macroscopic level. In order to justify MacroPT, he says, we would need either (1) independent reason to believe in the existence of macroscopic powers (which we would have, for example, if our reasons for believing in fundamental powers were also reasons for believing in macroscopic powers) or (2) reason to believe that the existence of fundamental powers would imply the existence of macroscopic powers (which we would have, for example, if we could show that fundamental powers could work together to constitute composite macroscopic powers). According to Bird, we have neither kind of justification for MacroPT.

Now, Bird means something slightly different from what I do by the term 'power'. According to Bird, the essential feature of a causal power—a feature that, he says, is agreed on in all the recent literature on the topic—is that causal powers are properties with a dispositional nature that is 'modally fixed'. The very same power, he says, 'could not have a different dispositional character or causal role: that character or role is fixed across possible worlds' (Bird, 2016, 6). So, for example, if negative charge is (amongst other things) the power to repel other negatively charged objects, then modal fixity says that it is not metaphysically possible to have the property of negative charge without being disposed to repel other negative charges. Modal fixity has not featured so far in my discussion of causal powers, though it will come up in Chapter 7.

For the rest of this section, let us distinguish powers$_{mf}$, which are powers in Bird's sense of being modally fixed, from powers$_i$, which are powers in my sense of being dispositions to manifest causal influence (note that these classifications are not exclusive: a power could be both). We can then distinguish two versions of MacroPT: MacroPT$_{mf}$, which refers to macro powers$_{mf}$, and MacroPT$_i$, which refers to powers$_i$. Bird is concerned with MacroPT$_{mf}$, but his challenge could equally apply to macro powers as I conceive them. In particular, it could be argued that we should not endorse MacroPT$_i$ unless (1) we have independent reason for believing in macroscopic powers$_i$ or (2) we can show that the existence of fundamental powers$_i$ implies the existence of macroscopic powers$_i$.

In fact, this challenge to MacroPT$_i$ is easily answered, since we have both kinds of justification for believing in in macroscopic powers$_i$. First, we have independent reason for believing in macro powers$_i$. For powers$_i$ are assumed by the reductive method of explanation, a method that has proved extremely successful. To the extent that this method is successful at the macroscopic level, therefore, we are justified in believing in the macroscopic powers$_i$ that the method assumes. Second, the existence of certain powers$_i$ at the fundamental level does indeed imply the existence of powers$_i$ at the macro level. For, as described above, whenever there are two or more fundamental powers$_i$ of the same type acting on the same object, there is a composite (and therefore macro) power$_i$ that is constituted by them.

There is no reason, then, to worry that Bird's objection to MacroPT$_{mf}$ will transfer across to an objection to MacroPT$_i$ and thereby raise a problem for my account of reductive explanation in macroscopic cases. But, in fact, the discussion in this chapter gives us the resources to defend even MacroPT$_{mf}$ from Bird's objections.

Bird's concern with MacroPT$_{mf}$ is that it cannot be justified in either of the two possible ways. We do not have independent reasons for believing in macro powers$_{mf}$, he says, because all the good arguments for modal fixity (arguments dealing with cross-world identity of properties and laws of nature) apply only to powers at the fundamental level. Furthermore, Bird argues, there is no good reason to believe that modal fixity at the fundamental level will imply modal fixity at the macro level. The best way to demonstrate such an implication would be to give an account of how macro powers$_{mf}$ are composed from fundamental powers$_{mf}$, and show that such composition preserves modal fixity. But, says Bird, we have no such account of how powers$_{mf}$ are composed.

It is on this last point that I think we can now resist Bird's argument, for in this chapter we have developed just what Bird says we need: an account of how powers can be composed. Of course, the account here developed is concerned with powers$_i$, not powers$_{mf}$, but it can still do the job Bird requires. For suppose that we accept Bird's argument that we have good reason to believe that fundamental causal powers are modally fixed. In particular, suppose that the fundamental powers referred to in reductive explanation are modally fixed, so that all fundamental powers$_i$ are powers$_{mf}$. Then, the account of composition of powers$_i$ will also be an account of composition for a large class of fundamental powers$_{mf}$. All that remains to be shown is that this composition will preserve modal fixity, so that the resultant macro powers are also powers$_{mf}$.

Suppose, then, that P_1 and P_2 are fundamental powers$_i$ to exert influences I_1 and I_2 on b respectively. Suppose also that P_1 and P_2 compose to form a composite power$_i$ P to exert an influence I on b. So $P = P_1 + P_2$ and $I = I_1 \oplus I_2$. Now suppose that P_1 and P_2 are modally fixed. This means that the functions I_1 and I_2 which describe these powers are fixed across all possible worlds. Now, by definition, the composite power$_i$ P is the power to manifest $I = I_1 \oplus I_2$. Since, by assumption, I_1 and I_2 are the same across all possible worlds, I will also be constant across all possible worlds if, and only if, the composition operator, $\oplus$, for I_1 and I_2 is also modally fixed. Thus, the composite power$_i$ P will also be a modally fixed power$_{mf}$ so long as the action of $\oplus$ is fixed across possible worlds. Now, in Chapter 7, I will argue that the composition laws for influences are grounded in the essential nature of these influences. If I am right about this, then the composition operator for I_1 and I_2 must be modally fixed, and so, therefore, is P.

In summary, we have good reason to believe in macro powers$_i$, and if we assume that fundamental powers$_i$ are modally fixed (and so are powers$_{mf}$), then it follows that macro powers$_i$ are also modally fixed, and so are macro powers$_{mf}$. All that remains in order to defend MacroPT$_{mf}$ is to defend the assumption that fundamental powers$_i$ are modally fixed. Bird defends modal fixity at the fundamental level by arguing that modally fixed powers are necessary to make sense of the laws of nature. In Chapter 7, I will argue that the we need modally fixed powers$_i$ to ground the laws of nature, and hence that Bird's arguments are in fact good reason to think that at least some fundamental powers$_i$ are powers$_{mf}$.

6.7 Are All Influences Forces?

The picture I have developed here is one in which causal influence is defined by a particular functional role. At the fundamental level, this role seems to be filled by forces, or whatever takes their place in quantum field theory. If a number of forces are acting on the same object at the same time, then this collection of forces taken as a whole can also fill the role of a force, and so we can treat the collection as a composite influence. But fundamental forces can also chain together asynchronically in complex ways to produce composite influences that are not directed towards accelerations and so are not themselves forces.

Consider, for example, how heat is transferred from one body to another through conduction. Each of the two bodies is composed of many smaller particles that are constantly jiggling around. The temperature of each body is just the mean kinetic energy of the vibrating particles that it is made of. When two bodies are brought together, the particles in one body can exert forces on the particles in the other by bumping into them (these forces are ultimately electromagnetic in nature). These interactions transfer energy between the particles by changing the particles' velocities. If a fast moving particle hits a slow one, the result is usually a transfer of energy from the faster particle to the slower. Now, suppose that one of our two bodies is warmer than the other. Then, the particles that make up the warmer body are, on average, moving faster than the particles in the cooler body. Collisions between particles in the two bodies, therefore, will tend to transfer energy from the warmer body to the cooler. Over time, the average result of all these collisions is that the particles in the warm body slow down and the particles in the cool body speed up, until the average kinetic energy in the two bodies is the same. Thus, the warm body cools and the cool body warms until they have the same temperature.

At the fundamental level, all the interactions involved in conductive heat transfer involve forces. However, the outcome of all these interactions is that the warm body has the power to exert a warming influence on the cooler body. This influence is not a force, since it is directed at a change in temperature, but it plays the role of an influence nonetheless. Warming influences from different bodies can be added together, and the influence that one body has on another depends only on the properties of the two bodies themselves, regardless of what else is going on. The theory of thermodynamics, which deals with the flow of heat rather than changes in the velocities of particles, takes advantage of the fact that fundamental forces can compose to form influences that are not themselves forces.

6.8 Conclusion

In this chapter I have argued that causal powers can, indeed, be composed of more fundamental powers. The argument for the compositionality of powers rests on the claim that causal influences can be composed of more fundamental causal influences, so much of the chapter was spent defending this claim from objections that have been put forward in the special case of forces. The fact that powers can compose to form more complex powers is important because it underlies the possibility of applying the reductive method at different 'levels'. If powers did not compose in this way, then we could only include truly fundamental powers in the explanatory basis of any reductive explanation, and most of our successful reductive explanations would not be possible. We could not explain population dynamics in terms of interactions between predator and prey species, we could not explain economic changes in terms of interactions between producers and consumers, and we could not explain the tides in terms of interactions between the oceans, the Earth, the Moon, and the Sun.

What is more, I have shown that the composition of causal powers preserves modal fixity, so that if there is a set of fundamental powers that are modally fixed, then any powers composed from these will also be modally fixed. Thus, we have a response to Bird's claim that there is no good reason to believe in the existence of macroscopic causal powers.

There is, however, a price to be paid when working with macroscopic causal powers. For such powers will tend to be of high rank, involving many basic objects and many variables. Such powers are difficult to deal with and, indeed, undercut the motivation behind the reductive method. Fortunately, there are situations in which many of these variables make very little difference and can be effectively ignored. In such cases we can work with a low-rank approximation to the true macroscopic power, and it is these approximations that make the reductive method useful in the macroscopic realm. I suggest that the various sciences are distinguished by their choices of explanatory bases, with each explanatory basis containing its own set of approximate causal powers.

Nonetheless, working with approximate causal powers has its drawbacks. Such powers may impose a limit on the accuracy of explanation or prediction, and they may only remain appropriately invariant within a given context. Thus, such approximations may have only limited applicability, and may be open to counterexamples. When explanations based on such powers break down, however, we do not need to abandon the reductive method, for it may be possible to find more accurate—or more invariant—powers by moving to a basis at a lower level. As we approach the fundamental level, we find that our low-rank descriptions of causal powers tend to be more accurate and more universally applicable. This suggests that the true fundamental powers are of low rank, each involving relatively few variables (or, if there is no fundamental level, it suggests that the powers below a certain level are all of low rank, each involving few variables). Of course, there is always the possibility that there are some truly fundamental high-rank powers. I will investigate that possibility in Chapter 10.

7
Laws of Nature

In science, and indeed in philosophy, we often find mention of 'laws of nature' such as Newton's law of gravitation and his three laws of motion, Coulomb's law of electrostatics, Mendel's laws of inheritance, the laws of supply and demand, and so on. But what, exactly, is it to *be* a law of nature? What kind of things are laws and why are they so useful in our attempts to understand the world? These are the basic questions that have occupied metaphysicians and philosophers of science who consider laws of nature, and in this chapter, I investigate the ways in which an ontology of power and influence might help us find answers.

Following the bulk of the philosophical literature, I take it that a law of nature is a statement which makes a general claim about the world; in the philosophical jargon, it is a universal generalization. So, for example, Newton's law of gravitation is the general claim that 'every massive particle in the universe attracts every other massive particle with a force which is directly proportional to the product of their masses and inversely proportional to the square of the distance between them'. Not all philosophers understand the term 'law of nature' in this way. Stephen Mumford (2004), for example, argues that—outside philosophy at least—the term 'laws of nature' is intended to refer to independent entities that govern the behaviour of objects in the world in such a way as to make the universal generalizations mentioned above true. However, as we shall see in a moment, the answer to the question of what it is to be a law of nature in the way I understand the term boils down to the question of what it is that makes the universal generalizations mentioned above true, and one way they could be made true is via the existence of laws in Mumford's sense. Thus, although I use the term 'law of nature' differently from Mumford, I think the basic metaphysical questions at stake are not so different.

If we agree to use the term 'law of nature' to refer to universal generalizations, then the first thing we note is that not all universal generalizations should count as laws. First of all, for a universal generalization to count as a law, it must be true; it is certainly not a law of nature that all humans can fly unaided.[1] But simple truth is not enough. It is true, for example, that every chair in my living room is brown, but surely this is not a law of nature; this truth is too contingent,

---

[1] Thus, we no longer think that Newton's law of gravitation is actually a law of nature. Scientists and engineers still use Newton's 'law', however, since in everyday situations things behave pretty much as if the 'law' were true.

too accidental. The problem is not simply that this statement about chairs in my living room is restricted to a particular spatio-temporal region, as there are true universal generalizations that are not restricted in this way, and that do not seem to count as laws of nature. Consider, for example, the statement that 'all solid spheres of gold have a diameter of less than one mile'. There are good reasons to believe that this universal statement is true (due to the scarcity of gold in the universe), but this seems to be an accidental feature of the world: a sufficiently advanced and dedicated civilization could, conceivably, gather or manufacture enough gold to create a gold sphere with a diameter greater than a mile. Compare the statement about gold with the statement that 'all solid spheres of enriched uranium have a diameter of less than one mile'. This latter generalization also seems to be true, but it seems to be non-accidentally true since uranium's critical mass ensures that it is not possible for a sphere of enriched uranium this size to exist (van Fraassen, 1989, 27). A law of nature, therefore, is a *non-accidentally true* universal generalization, and the question of what it is to be a law of nature boils down to the question of what distinguishes the accidentally true universal generalizations from the non-accidentally true ones.

Over the past fifty years or so, there have been two main answers to the question of what distinguishes a law from an accidentally true universal generalization: the *systems* view, and the *universals* view. The systems view of laws is based on ideas from Mill (1843) and Ramsey (1978), and has been defended by Lewis (1973b; 1983a; 1994a), Earman (1984), and Loewer (1996). The basic idea of the systems view is that a law of nature is distinguished by the fact that it will be a part of all the true deductive systems that best balance strength and simplicity. So, laws of nature are statements that are particularly efficient at describing features of the world. The systems view is motivated by a Humean metaphysics, according to which there are no necessary connections between distinct entities. In particular, any object could have any combination of fundamental properties and the properties of one object do not restrict the properties that can be had by another wholly distinct object. According to the Humean, therefore, there is nothing other than the pattern of events in the world that can make a universal generalization true, and this is why we must turn to logical features of systems of statements to distinguish laws from non-laws.

The universals view of laws has been championed by Tooley (1977), Dretske (1977), and Armstrong (1983). According to the universals view, a universal generalization is a law of nature if it is made true by a relation of non-logical necessitation between the universals mentioned in the law. So, for example, the statement that 'all Fs are Gs' is a law of nature if it is made true by the fact that there is a relation of non-logical necessitation between F and G, written $N(F, G)$. The universals view denies the Humean ontology, since it posits necessary connections between universals. However, as Alexander Bird (2007, 2) notes, the view is semi–Humean, since it is a *contingent* fact that F and G are related by N. There is nothing in the nature of the properties F and G themselves that ensures they are related by N, so any universal could be related to any other by this relation of non-logical necessitation.

In the past few decades, however, a third, *dispositional essentialist* approach to the laws of nature has been developed. The basic idea, which has been championed by Swoyer (1982), Bostock (2001), Kistler (2002), and Bird (2007), is that the laws of nature are those universal generalizations that are made true by the dispositional essences of the properties involved. This approach to laws of nature clearly assumes a non–Humean ontology, since it involves properties with dispositional essences, which is to say that it involves properties whose essences are tied to manifestations that may include other properties. So if there are dispositional essences, then there are necessary connections between properties. What is more, these necessary connections are not contingent: they are part of the essence of the properties involved.

My aim in this chapter is not to compare these three approaches to the laws of nature. Bird (2007) has already done a fine job of that, arguing persuasively that the dispositional essentialist account is preferable to the other two. The ontology of power and influence seems to be a good fit with the dispositional essentialist view of laws, and so my aim in this chapter is to investigate how this ontology might help develop the dispositional essentialist view. I will focus in particular on the derivation of laws from dispositions as it is presented in Bird's book *Nature's Metaphysics: Laws and Properties* (2007), since I think this is the most developed account of the dispositional essentialist view to date. I will argue that Bird's derivation only works if we combine the view that some properties are powers in Bird's sense (they have an essential dispositional nature that is modally fixed) with the view that the relevant properties are powers in my sense (they are dispositions to manifest causal influence).

7.1 From Dispositions to Laws

As discussed in Chapter 4, there is a deep connection between dispositions and subjunctive conditionals. This thought underlies the simple conditional analysis of dispositions:[2]

(SCA) x is disposed to manifest M in response to stimulus S iff were x to undergo S, x would yield manifestation M.

---

[2] I follow Bird in considering only single-track dispositions described by counterfactual conditionals. However the argument is unchanged if we consider multi-track dispositions, so that M is a function of S.

Now, if we let $D_{(S,M)}$ represent the disposition to manifest M under stimulus S, then the simple conditional analysis can be written as follows:

(CA) $D_{(S,M)}x \leftrightarrow (Sx \;\Box\!\!\to M x)$.

Since (CA) is supposed to be a conceptual analysis of what it means to have the disposition $D_{(S,M)}$, Bird argues that if (CA) is true, it is necessarily true. Thus, the disposition $D_{(S,M)}$ can be characterized as follows:

(CA$_\Box$) $\Box(D_{(S,M)}x \leftrightarrow (Sx \;\Box\!\!\to M x))$.

Suppose now that P is some essentially dispositional property, then to possess P is to possess a disposition to yield some manifestation M in response to some stimulus S. Thus, for any object x,

(DE$_P$) $\Box(Px \to D_{(S,M)}x)$.

I will discuss the necessity operator in a moment. Finally, (CA$_\Box$) and (DE$_p$) together imply

(V) $\forall x((Px \;\&\; Sx) \to Mx)$

Now, (V) is a universal generalization. What is more, this generalization does not describe some merely accidental regularity, since the truth of (V) is grounded in the nature of the property P and the necessary truth (CA$_\Box$). Thus, there is good reason to view (V) as a law of nature. This derivation of laws from the dispositional essence of a property, says Bird, is 'the core of the dispositional essentialist explanation of laws' (2007, 46).

It should be clear at this point that, like me, Bird is taking 'law of nature' to refer a particular kind of statement (a non-accidentally true universal generalization) rather than the truth maker for such a statement. Recall, however, that Mumford (2004) takes 'laws of nature' to refer to real and distinct existents which somehow govern or control events and thereby explain the regularities we see in nature. Thus, although Mumford advocates an ontology of powers very like the one that is assumed here, and even though he would presumably agree that there are non-accidentally true universal generalizations, he argues that there are no laws of nature. For, says Mumford, the regularities we see are explained by powers simply doing what they do; these powers are not governed or controlled by some other entity that could count as a law of nature. I take it that the only real disagreement between the view of laws that Bird (and I) advocates and the view defended by Mumford, is about what gets to be called a 'law'.

It is interesting to take note of the role that modal fixity plays in Bird's derivation. The claim that P has a real dispositional essence that is fixed across all possible worlds is captured by the necessity operator in (DE$_P$). However, if we remove this operator, and so replace (DE$_P$) with

(D$_P$) $(Px \rightarrow D_{(S,M)}x)$,

then we can still derive the universal generalization (V). So one could be forgiven for wondering whether the assumption of modal fixity plays any role at all. The answer is that despite the derivation, without modal fixity we do not get a satisfying *explanation* of the generalization (V). Since (V) relies on (D$_P$), a satisfying explanation of (V) requires an explanation for the truth of (D$_P$). If P has a real dispositional essence, then we can explain the truth of (D$_P$) as a necessary consequence of this essence—as described by (DE$_P$). If, on the other hand, the dispositional nature of P is not modally fixed, meaning that P does not have a dispositional essence, then we will be at a loss to explain the truth of (D$_P$). All we can say is that objects with property P have the disposition $D_{(S,M)}$, but they could have had some other disposition. What we have, in this case, is merely a description of a regularity, rather than an explanation. The whole point of dispositional essentialism is to explain, rather than merely describe, the regularities that we observe in the world and this cannot be done without modal fixity.[3]

The dispositional essentialist account of laws is lovely, but it faces a problem. As discussed in Chapter 4, (SCA), and hence (CA$_\square$), are generally regarded as false because of the possibility of finks and antidotes. An object's disposition is said to be *finkish*, remember, if the stimulus that would trigger the disposition also causes the object to lose that disposition, and lose it before the manifestation has had time to come about. An *antidote* (also known as a *mask*) does not remove a disposition, but, rather, interferes with the causal chain leading from the triggering of a disposition to its manifestation. Either case counts as a counterexample to the counterfactual conditional within (CA$_\square$), since in both cases the antecedent is satisfied, while the consequent is false. Thus, (CA$_\square$) is false.

One might think that the falsity of (CA$_\square$) deals the dispositional essentialist account of laws a fatal blow. However, Bird (2007, 60) argues that the falsity of (CA$_\square$) is actually an advantage for dispositional essentialism, since we get an even more powerful theory if we replace (CA$_\square$) with a more accurate analysis. In particular, in order to make allowances for finks and antidotes we can replace the left-to-right half of the conditional analysis with:

(CA$\rightarrow$*) If x has the disposition $D_{(S,M)}$ then, if x were subject to S and finks and antidotes to $D_{(S,M)}$ were absent, x would manifest M.

---

[3] These thoughts do, however, suggest a novel Humean view of laws. One thing that an adequate account of laws needs to do is distinguish true laws of nature from merely accidentally true universal generalizations. The standard Humean approach (as advocated by Lewis, 1973b) is to distinguish laws as those true universal generalizations that figure in the 'best system' for describing the world. The alternative suggested here is for the Humean to agree with the dispositional essentialist that the laws of nature are those generalizations that are guaranteed by the dispositional behaviour that is associated with one or more of the constituent properties—in the way that (V) is guaranteed by (D$_P$)—whilst denying (DE$_P$). The Humean, note, can accept truths like (D$_P$) since they posit only contingent relationships between properties and dispositions.

Plugging this new version of $(CA_\square)$ into the derivation above gives us the following instead of (V):

(V*) $\forall x$(finks and antidotes to $D_{(S,M)}$ are absent $\rightarrow ((Px \ \& \ Sx) \rightarrow Mx))$

Now (V*) is a *ceteris paribus* law; it describes a regularity that admits of exceptions. So the dispositional account of laws, it seems, can be used to explain both strict, exceptionless laws and *ceteris paribus* laws. The falsity of the simple conditional analysis actually give the dispositional essentialist account of laws *more* explanatory power rather than less.

7.2   Fundamental Finks and Antidotes

Bird defends the dispositional essentialist account of laws by showing that it can explain both strict laws and *ceteris paribus* laws. However, more needs to be said in order to make this defence work. In particular, we need to show there are at least some circumstances in which dispositional properties are not plagued by finks or antidotes. To see this, suppose that the manifestation of some dispositional property, P, is finkish, or is interrupted by an antidote, in almost all circumstances. In this case, P would almost never get a chance to show itself, and would almost never actually play any role in what comes about. Thus, P could not ground any regularity in events, and so could not explain any law of nature. In general, the dispositional essentialist derivation above will only give rise to observable regularities in situations where there are few, if any, finks or antidotes. If there are no such situations, then no regularities will be the result of dispositional essences and dispositional essences cannot underlie the laws of nature.

So, are there any situations in which finks and antidotes are rare? Bird argues that there are (though he does not seem to appreciate the importance of establishing this, given my argument in the previous paragraph). In particular, Bird argues that at the fundamental level of reality (assuming there is such a level) finks are not possible and antidotes are at least unlikely (2004; 2007, 60–3). I find Bird's argument against fundamental finks convincing and will not go into the details here. His argument against fundamental antidotes, however, is not so clearly correct, so it is worth further investigation.

Examples of antidotes at the macroscopic level all work by interrupting a chain of processes, or mechanism, that leads from the triggering of the disposition to its characteristic manifestation. When a fragile glass is struck, for example, it does not shatter immediately. Rather, a pattern of shock waves spreads across the glass from the point that it is struck, and the stresses produced by these waves at each point cause the glass to fracture. A brilliant sorcerer (or physicist), therefore, could provide an antidote to the glass's fragility by applying shock waves that perfectly cancel the waves produced by the striking of the glass, thus saving the glass from destruction without removing its intrinsic fragile disposition (Bird, 2007, 28). But, according to Bird's understanding of the term 'fundamental', a disposition is not fundamental if it gives rise to its manifestation via a chain of intermediary processes (in the language of Chapter 6, a disposition that is the resultant of asynchronic composition cannot be fundamental). Thus, Bird points out, an antidote cannot affect a fundamental disposition in the usual way, namely, by disrupting the chain, or mechanism, which leads from the disposition to its manifestation.

The only way for antidotes to arise at the fundamental level, Bird concludes, is what he calls possibility (b):

(b) That there is no mechanism bringing about the manifestation M—it is brought about by D [the disposition] and S [the stimulus] together directly. However, the further possible condition A is such that D and S will not bring about M.   (Bird, 2007, 63)

Possibility (b) seems rather odd, but Bird concedes that it is an empirical question whether there are situations that realize this possibility. However, he tells us that the 'direction of the development of physics with ever fewer fundamental properties and corresponding forces indicates that the prospects for antidote-free fundamental properties and thus strict laws only at the fundamental level are promising' (Bird, 2007, 63).

Nancy Cartwright, however, sees things differently. According to Cartwright, all laws, even the fundamental laws of physics, are *ceteris paribus* laws. In particular, we can read Cartwright to be claiming that antidotes are not rare even at the fundamental level. Thus, Cartwright's claim that all laws are *ceteris paribus* laws poses a challenge to the dispositional essentialist derivation of laws. Let us turn to that challenge now.

7.3 Lying Laws: Regularities in the Course of Events

Consider a situation in which two similarly charged objects are in close proximity to each other. According to Coulomb's law, each object will feel a force directed away from the other. Now Newton's second law, $F = ma$, tells us that when an object is subject to a force, the object will accelerate in the direction the force is pointing.[4] Together, then, Coulomb's law and Newton's second law imply that two similarly charged objects will accelerate away from each other.[5] In the real world,

---

[4] More precisely, given the discussion of Section 5.5.2, if an object is subject to a force F, and F is directed towards an acceleration of a m/s in direction $\mathbf{u}$, then the object will undergo an acceleration of magnitude a and direction $\mathbf{u}$.

[5] In drawing this conclusion we are, of course, failing to distinguish total and component forces. I will address this issue below.

however, charged objects do not often behave this way. The postively charged protons within atomic nuclei remain bound together rather than accelerating away from each other; positively charged polystyrene balls can be pushed together with ease; and Cartwright has described a set-up in which two similarly charged particles will actually move closer together as a result of the coulomb force between them (1999, 60–1). Considerations such as these lead Cartwright (1983) to provocatively claim that the laws of physics lie.

Put less provocatively, Cartwright's point is that the kinds of regularity that are described by the laws of nature are extremely rare. The mere fact that two charged objects are in proximity to each other tells us nothing about what those objects will actually do. They could move apart, move together, remain stationary, or exhibit some other more complex behaviour, depending on the environment in which they find themselves. Since the environments in which charged objects can find themselves are many and varied, the behaviours of such objects will also be many and varied. In order to observe the regularities described by laws like Coulomb's law and Newton's laws, Cartwright notes, we must go to a lot of trouble to screen off any outside interference.

At this point we should note that Cartwright's conclusion is stated more strongly than is warranted. The discussion of Coulomb's law above, for example, only shows that we should not expect regularities in the *motions* of charged objects. It does not follow that there can be no regularities among other properties or entities (more on this below). In general, Cartwright's arguments only show that regularities are rare in what she terms the 'course of events'. That is, we should not expect regularities among the kinds of entities and properties that are acceptable in a Humean metaphysics: objects with properties such as position, size, colour, velocity, mass, charge, etc. But if we restrict our attention to the course of events, Cartwright's conclusion that regularities are very rare seems correct. Even two of Cartwright's greatest critics, John Earman and John Roberts, accept Cartwright's claim when restricted to 'those branches of physics that aim to state correlations among more or less observable macroscopic phenomena' (1999, 473).

Consider, then, how Cartwright's conclusions affect the dispositional essentialist's understanding of the laws of nature. If we assume that the regularities produced by dispositional essences are regularities in the course of events, then Cartwright's arguments pose a direct threat to dispositional essentialism, for Cartwright has shown that such regularities are very rare. If we restrict our attention to regularites in the course of events, the only way to reconcile the dispositional essentialist derivation of laws with Cartwright's arguments against such regularities is if the antecedents of one or more of the left-to-right conditionals involved in the dispositional essentialist derivation are rarely satisfied.

Suppose then that the left-to-right antecedents of $(CA_\square)$ and/or (DE_p) are rarely satisfied for some property P with dispositional essence D. What this means is simply that the property P is rarely instantiated. But if a property P is rarely instantiated, then observed behaviours must usually be the result of properties other than P, or not be the result of any properties at all. And in order to secure the rarity of true regularities in this way, the same must be said about all properties with dispositional essences. So, if the left-to-right antecedents of $(CA_\Box)$ and/or (DE_P) are rarely satisfied, then either most behaviours are not the result of a property with a dispositional essence, or the large variety of behaviours are matched by a similarly large variety of dispositional properties, one for almost every possible situation in which an object may find itself. Neither of these options should seem very palatable to the dispositional essentialist.

Similarly, if the antecedents of (V) are rarely instantiated, then either property P is rarely instantiated or the triggering conditions of P are rarely instantiated. In this case, we are led to the conclusion that either most behaviour is not the result of the triggering of a dispositional property or the large variety of behaviours is matched by a similarly large variety of dispositional properties. Once again, neither result is very palatable to the dispositional essentialist.

The only other possibility for reconciling dispositional essentialism with Cartwright's conclusion is to suppose that it is the antecedent of (V^*) that is rarely instantiated. But, to suppose that the antecedent of (V^*) is rarely instantiated is to suppose it rare that finks and antidotes to any disposition are absent. This, I believe, is the situation that Cartwright thinks is the case, for the kinds of situations that Cartwright describes in which the regularities described by a law are not manifest are all situations in which there is some outside interference. The disposition that two protons in an atomic nucleus have to accelerate away from each other in virtue of their positive charge, for example, does not manifest because it is interfered with by the strong nuclear force between them. Such interference seems to match precisely Bird's possibility (b): an object (the proton, say) has a dispositional property (positive charge) which is triggered (by proximity to another proton), but the characteristic manifestation (acceleration) is prevented due to the interference of some outside factor (the strong nuclear force).

What Cartwright's arguments suggest, therefore, is that dispositional properties, even fundamental properties like electric charge, will almost always be plagued by antidotes (so long as we assume that the manifestations of dispositions occur in the course of events). But we have seen in the previous section that dispositional essentialism requires that there are some situations in which finks and antidotes are unusual, if not completely absent. Thus, Cartwright's arguments pose a direct threat to the dispositional essentialist account of laws.

Now, Cartwright does allow that there are situations in which regularities can be observed in the course of events. Typically, these are laboratory conditions in which variables are controlled and the dispositions in question are shielded from outside interference to produce what Cartwright calls a 'nomological machine' (1999, 50). But if dispositional properties only give rise to regularities within a laboratory context, then one wonders how it is that learning about these properties can be useful in the outside world. How is it that we can use this knowledge to build bridges, launch spacecraft, or decide not to let our children play in heavy traffic? It is not enough that dispositions be free from antidotes in some very restricted class of situations: they must be free from antidotes in the situations in which we successfully put our knowledge of these dispositions to good use.

In summary, if we assume that dispositional manifestations only occur within the course of events, then the dispositional essentialist derivation of regularities conflicts with Cartwright's argument that there are no regularities within the course of events. So if the dispositional essentialist wants to claim that dispositional properties ground the laws of nature, then she must allow that the laws of nature, and hence the dispositions that ground them, involve something that goes beyond the course of events.

7.4 Laws of Influence

We can avoid Cartwright's arguments by claiming that the regularities described by laws of nature involve something beyond the course of events. The usual culprits for this 'something extra' are component forces.[6] Proponents of this response to Cartwright include Creary (1981), Forster (1988a; 1988b), Earman (1999), and Smith (2002). Indeed, I too have argued that component forces will do the job (Corry, 2006). Component forces, of course, are a species of causal influence, so naturally I will suggest that if the laws of nature are to be grounded in essentially dispositional properties, then these dispositional properties must involve causal influences. In particular, Cartwright's arguments can be avoided if the dispositions involved are dispositions to manifest influences. That is to say, they are causal powers in the sense that I defined them in Chapter 4.

Recall that fully specified fundamental powers are immune to finks and antidotes. The influence that a power manifests depends entirely upon the state of the system associated with that power, and so is immune to changes in properties outside this system. The law describing the component force between a pair of charged particles, for example, only takes into account properties of, and relations between, the two particles themselves. It is thus irrelevant whether or not the pair are part of a more complex system.

Suppose, then, that P is a fundamental power. Since fundamental powers are immune to finks and antidotes, (DE$_P$) and (CA$_\square$) will both be true of P, and so P will give rise to the strict universal generalization (V). So, as hoped,

---

[6] The total force on a particle is also something in addition to the course of events, and so might also be thought to be a candidate. But since total forces can be tied one-to-one with actual accelerations within the course of events, regularities involving total forces will be as rare as regularities in the course of events.

fundamental powers will give rise to strict universal generalizations. Since these generalizations are grounded in the essential dispositional nature of the powers involved, they will not be merely accidental truths and so are worthy of the title 'law of nature'. To allow this possibility, of course, we have had to allow that the regularity described by these generalizations involve entities that go beyond the course of events. In particular, the manifestation, M, that appears in (V), is a causal influence. Since laws of this form tell us which influences are present in any given situation, I will call them *laws of influence* (this terminology was first introduced by Creary 1981).

For example, according to the metaphysics of power and influence, the manifestation of electric charge is a force not an acceleration. Thus, if we plug the appropriate definition of electric charge into (DE_P), the resulting proposition (V) will be something like 'whenever an object with charge q_1 is a distance r from another object of charge q_2 there will be a force $f = \frac{\kappa_e q_1 q_2}{r^2}$ on the second particle (where κ_e is a constant)'. This is just Coulomb's law. The fact that there are almost always other forces at play means that the second particle typically will not accelerate at the rate f/m, so its behaviour in the course of events will not be regular. But Coulomb's law nonetheless describes a true regularity in the manifestation of forces.

What I have shown, then, is that dispositional essences can ground laws of nature, so long as the dispositions involved are dispositions to manifest influences, and the laws involved are laws of influence. But not all laws of nature are laws of influence, and we might legitimately wonder what grounds these other laws. For example, the dispositional essentialist might tell us that Coulomb's law is the result of the dispositions that charged particles possess to produce component forces. But what of Newton's second law, or the force composition law which tells us that the total force is the vector sum of the component forces? What, if anything, grounds these regularities? In a previous paper (Corry, 2011) I argued that this this question poses a serious problem for the dispositional essentialist. However, the understanding of causal powers and causal influence that I have developed in this book suggests an answer to this problem, and I turn to that answer now.

7.5 Laws of Composition

In Chapter 6, I argued that the characteristic outcome towards which a force is directed is an essential part of what it is to be that force. To be a force of magnitude f and direction $\mathbf{u}$ is to bring about an acceleration of magnitude f/m and direction $\mathbf{u}$ when acting in isolation on an object of mass m. If this is correct, then if we restrict our attention to forces that act in isolation, Newton's second law, $f = ma$, follows from the essential nature of forces. But what about cases in which forces are not acting in isolation? When there is more than one force acting on an object, it is unlikely that the acceleration that occurs is the characteristic acceleration of any of the individual forces. In general, a component force of magnitude f is unlikely to produce an acceleration of magnitude f/m when acting on a mass m.[7]

In order to predict, or explain, what goes on in the course of events, it is not enough to know the component influences that are manifested at any given time, even if we know the characteristic acceleration that each influence would produce in isolation. We must also know what happens when these influences act together. Now, as discussed in Chapter 6, component influences acting together instantiate resultant influences. Like component influences, then, a resultant influence will have a characteristic outcome towards which it is directed. In particular, if a resultant force has magnitude f, direction $\mathbf{u}$, and acts on an object of mass m, then it will be directed towards an acceleration of magnitude f/m in direction $\mathbf{u}$. What is more, if f is the resultant of *all* the forces acting on m, then, according to assumptions 3 and 4, the outcome towards which it is directed will actually occur (since there are no other relevant influences acting on m). So, the outcome that occurs will be the one that is predicted by Newton's second law as applied to the total resultant force. Thus, we can understand the second law as a result of the essentially directed nature of forces, so long as the law is restricted to the total resultant force in the situation (or at least an appropriate approximation to the total resultant force).

The picture I have painted so far is one in which dispositional essences ground laws of influence which, like Coulomb's law or Newton's law of universal gravitation, tell us all the component influences that are present in a given situation. The actual outcome that occurs is then the characteristic outcome towards which the resultant influence is directed, and it is this fact that grounds Newton's second law for forces. But there is still something missing: in order to apply the laws discussed so far we need to know what resultant influence will be produced by a given set of components. In addition to laws of influence, then, we need *laws of composition* which, as in the case of the vector addition law for forces, tell us how to combine component influences.

Laws of composition cannot be grounded in the dispositional essences of properties, since these dispositions only give rise to regularities of the form (V), and (V) does not tell us how to add influences. So, where do these laws of composition come from? Must we conclude, with the Humean, that the laws describe mere regularities that just happen to hold, there being no further explanation as to why they hold? I do not think so. I believe that the laws of composition could, like Newton's second law, be grounded in the essential natures of the influences to which these laws apply. In Chapter 5, I noted that the composition of influences forms an algebra. Suppose, then, that this algebraic structure is part of the essential nature of the influences involved. In other words, suppose that a given influence's

---

[7] Smith (2002, 253–4) makes much the same observation, and concludes that forces cannot be dispositions.

position in this algebraic structure is an essential part of the nature of that influence. In this case, the algebraic structure, and hence the laws of composition, would be grounded in the essential natures of the influences involved. My suggestion here is analogous to a similar suggestion for understanding the integers. It seems plausible that the position that an integer holds in the group structure of integers under addition is part of the essence of what that number is. Zero, for example, just is that number which, when added to any other number n, results in n, and it is part of the essential nature of the integers 1 and 2 that $1 + 1 = 2$. Similarly, I suggest, it is part of the essential nature of any force f that it composes with other forces according to the algebra of vector addition.

If this suggestion is correct, then the characteristic outcome towards which an influence is directed is only part of the nature of that influence. A second— and logically independent—part of the nature of an influence is the position the influence holds in the algebra of composition. This picture is actually in perfect agreement with the discussion in Chapter 3 where I pointed out that any given causal influence can be characterized by two separate notions of magnitude. One magnitude describes the size of the change towards which the influence is directed; the other magnitude has to do with how it interacts with other influences, and so is related to the position the influence holds in the algebra of composition. My suggestion here is that both magnitudes contribute to the essence of an influence.

7.6 Macroscopic Powers and *Ceteris Paribus* Laws

I have argued above that fundamental powers and influences (if there are any) will give rise to regularities that can be appropriately described as laws of nature. In Chapter 6, however, I argued for the existence of non-fundamental macroscopic powers and influences. If these macroscopic powers are fully specified, and only involve synchronic composition (so that there are no chains of influence involved between each power and the final influence that it manifests), then they will be immune to finks and antidotes, and so they will also give rise to exceptionless regularities that are appropriately described as laws of nature. Like the composite powers that give rise to them, these laws are likely to involve numerous objects and numerous variables, and so be of limited practical use. An example of such a law would be one which states that if a bunch of particles of a particular type are arranged in a particular way, then the first of these particles will be subject to a particular force. Such a law will only be of much interest if we commonly find particles of this type and this arrangement.

In Chapter 6, I suggested that the most useful macroscopic powers are likely to be partly specified 'approximate powers' and will often involve asynchronic composition. In general, macroscopic powers that involve asynchronic composition, and powers that are only described approximately, will be open to finks and antidotes, and so will not give rise to regularities that can be described by strict universal generalizations. Nonetheless, such powers are useful because there are contexts in which they are sufficiently invariant to serve in reductive explanations. These useful macroscopic powers will also give rise to regularities, but these regularities may only hold approximately, and only within the appropriate context. Thus, these macroscopic powers will give rise to *ceteris paribus* laws and rules of thumb, rather than universal laws.

7.7 Conclusion

Alexander Bird has argued that laws of nature can be grounded in the essentially dispositional nature of causal powers. In this chapter I have shown that Bird's claim is correct, with some qualifications. First, in order to ground laws of nature it is not enough that the dispositions involved be causal powers in Bird's sense of being modally fixed. The relevant dispositions must also be causal powers in my sense; they must be dispositions to manifest causal influences. Second, it is only laws of influence that can be grounded by powers in this way. In particular, laws of composition cannot be so grounded. However, I have suggested that laws of composition can be grounded in the essential nature of the influences involved. Thus, an ontology of power and influence can provide a reasonably complete grounding for the laws of nature.

I began this chapter with reference to the Humean world view, according to which there are no necessary connections between properties or events. The world, according to this view, is just a vast mosaic of particular matters of fact. Laws of nature describe patterns in this mosaic but they are not part of the explanation for why these patterns occur, for according to the Humean there is no such explanation. It is as if the tiles were thrown down at random and it is only after the fact that we notice that there are some patterns in the way they just happened to fall. The non-Humean finds this lack of explanation unsatisfying. Douglas Kutach puts the worry as follows:

> Look around at the world our scientists have codified. To deny that there are laws at work and claim that there are just events here and events there is tantamount to claiming a massive coincidence. The Humean mosaic just so happens to be in complete accordance with the fantastically precise equations of physics without any exception. This is a textbook case of postulating a fantastic conspiracy.
> (Kutach, 2014, Ch. 2)

On the other hand, says Kutach, Humeans can respond as follows:

> You claim that there are some fundamental laws of physics *making* the world develop a certain way, but that is just taking a metaphor too seriously. Laws of

nature are not the kind of things that can push and pull material objects around or summon the future into existence. No matter how impressive the hoped-for laws of physics turn out to be, they are fundamentally just equations that express how various parts of the actual world are related to one another. Equations are not the kind of thing that can engage in causation. What's more, calling them laws does not reduce the mystery. It just re-packages the mystery under a new name. It doesn't explain the data any more than the hypothesis that the universe is the way it is because it was the will of God. Your appeal to laws is no more explanatory than postulating an undetectable ghostly spirit.   (Kutach, 2014, Ch. 2)

The view of laws advocated in this chapter satisfies the non-Humean desire for explanation of the regularities we see in the world, whilst at the same time avoiding the Humean objection above. We can agree with the Humean that the laws of nature are descriptions of the regularities that occur in the world, and are not what creates these regularities. Nonetheless, the regularities we see in the world are no mere coincidence: they are a result of the operation of the causal powers and influences that are instantiated in the world.

8

Causation

Throughout this book I have been using causally loaded language. I have spoken of 'causal powers' and 'causal influences', and have described these influences as 'acting on', or being 'exerted by', objects. The reason for this choice of language is that I feel that there is a deep connection between reductive explanation and causal explanation. Indeed, I believe that our causal concepts arise from, and find their most important use in, our attempts to explain the world in terms of the action of localized parts. In this chapter and in the following, I will defend this claim by showing that causal influences underlie our concept of causation. Before I can embark on this task, however, a little metaphilosophy is necessary, for, as Phil Dowe (2000), Douglas Kutach (2014), and others have pointed out, there are distinct tasks that philosophers may be taking on when they set out to give a theory of causation. Thus, I will begin with a few words to clarify the task that I am trying to accomplish here, and the approach that I will take to accomplishing it.

The first of the two tasks that Dowe identifies is *conceptual analysis*, which aims to elucidate our normal concept of causation. Conceptual analysis can be purely descriptive, identifying how ordinary people actually use the concept, or it can be revisionary, proposing changes to the concept based on considerations such as consistency and simplicity. Either way, the ideal aim of conceptual analysis is a set of necessary and sufficient conditions for the correct application of the concept. The data against which such an analysis is tested are the (considered) intuitions of competent users of the concept. As such, conceptual analysis typically proceeds a priori, making heavy use of thought experiments that, it is assumed, will elicit similar intuitions in all competent users of the concept (though, recently, the 'experimental philosophy' movement has started gathering statistical information about the intuitions elicited by various thought experiments, paying particular attention to the way these intuitions differ between different groups of people. See, for example, Machery 2011).

The second task that Dowe identifies is what he calls *empirical analysis*. Empirical analysis, he says, is concerned with what causation *is* in the world; it 'aims to map the objective world, not our concepts' (2000, 3). As with conceptual analysis, the ideal goal of empirical analysis is a set of necessary and sufficient conditions for the correct application of the concept. Unlike conceptual analysis, however, an empirical analysis is not tested against the intuitions of competent users: rather, it is tested against our best scientific theories. As a scientific realist, Dowe believes that scientific theories are the best guide we have to the structure of reality. Thus, in order to find out what causation is in the world, we should look to the role that the concept of causation plays in our best scientific theories. Since our best scientific theories are discovered a posteriori, Dowe believes that empirical analysis too can only proceed a posteriori.

Like Dowe, I will be concerned in this chapter, and the next, with the question of what, if anything, causation *is* in the world. However, I do not think we can simply look to our best scientific theories and read off the nature of causation. Science is not a monolithic enterprise, so to which branch should we look to tell us what causation is? Economics? Epidemiology? Climatology? Chemistry? Can we assume that all these sciences mean the same thing by 'causation'? Dowe looks to physics for the true nature of causation, and this might seem reasonable, because physics aims to investigate the fundamental components of reality. But, as Russell (1913) pointed out, the equations of fundamental physics make no explicit mention of causation. Following up Russell's insight, a number of philosophers have recently argued that the concept of causation simply does not apply at the fundamental level (see, for example Norton, 2007; Hitchcock, 2007). I am not so pessimistic about finding causation in physics, but I do feel that we should not be expecting to learn all we need to know about causation from physics alone. If we are to understand the role that causation plays in science, I contend, we should not focus on a particular branch of science. Rather, we should look for something that is common to the concept of causation employed across all branches of science.

The basic objects and properties involved in physics are somewhat different from those involved in chemistry, and they are both worlds apart from the objects and properties involved in economics, and so one might despair of finding *any* concept that all sciences have in common, let alone a concept of causation. However, there is something that all sciences, and, indeed, most systematic attempts to understand the world, have in common: they all make use of the reductive method of explanation. What is more, I have shown that the reductive method makes certain assumptions about the world: it assumes that there are causal powers which manifest influences which, in combination, change the world. It is in these powers and influences that I believe we will find the answer to the question of what causation is in the world. Thus, I suggest that we should not look to our best scientific *theories* to tell us the nature of causation: rather, we should look to the assumptions that underlie scientific *methodology*.

Uncovering the assumptions that underlie a methodology is a process that proceeds a priori. The discussion of the reductive method in the preceding chapters may be informed by empirical examples, but it was nonetheless an a priori investigation of the necessary conditions for the application of the method. The only empirical question is whether this method actually turns out to be success-ful in understanding the world. So, like empirical analysis, my aim here is to investigate what causation is in the world. Unlike empirical analysis, however, my investigation will proceed largely a priori. Thus, my approach here is not empirical analysis as Dowe defines it. Since my approach looks for causation in the necessary conditions for a certain way of understanding the world, let us call it *transcendental analysis*.

Although empirical analysis and transcendental analysis are both distinct from conceptual analysis, both approaches presuppose some kind of grasp of the ordinary concept of causation. As David Lewis says with regard to the philosophy of mind:

> Arbiters of fashion proclaim that [conceptual] analysis is out of date. Yet without it, I see no possible way to establish that any feature of the world does or does not deserve a name drawn from our traditional mental vocabulary.   (1994b, 415)

Similarly, we cannot ask what, in the world, corresponds to our concept of causation if we have no idea what that concept is. What is more, it may be that simply analysing our concepts can tell us something about the world. This would be the case, for example, if our concepts correctly latch onto natural kinds in the world, and so 'carve nature at the joints'. In this case, learning about the boundaries of our concepts can teach us about the natural kinds in the world. For these reasons I will begin by considering conceptual analyses of causation. However, my conclusion will be somewhat negative. I do not believe that the concept of causation can be given a univocal analysis in terms of necessary and sufficient conditions. Nor do I believe that the concept is likely to carve nature at the joints. Nonetheless, a consideration of attempts to provide conceptual analyses of causation will, I believe, justify my claim that powers, and the influences they manifest, deserve the title 'causal', and will suggest that our uses of the term 'causation' are grounded in the existence of causal influences.

8.1 The Problems of Analysis

Conceptual analysis of causation has been dominated by attempts to provide necessary and sufficient conditions for the correct application of the relation *c caused e*. The usual aim is a reductive analysis, which would provide a set of non-causal conditions that are necessary and sufficient for the obtaining of the causal relation. So, for example, regularity analyses are based on the idea that *c* caused *e* just in case *c*s are regularly followed by *e*s (see, for example, Mackie, 1974; Baumgartner, 2008); probability analyses are based on the idea that *c* caused *e* just in case the probability of *e* given *c* is greater than the probability of *e* given not-*c* (see, for example, Suppes, 1970; Eells, 1991); and counterfactual analyses are based on the idea that *c* caused *e* just in case *e* would not have occurred had *c* not occurred (for example, Lewis, 1973a). Such analyses are deemed successful in proportion to how well they agree with intuitive judgements about whether it is correct to assert that c caused e in any given case.

Anybody familiar with the literature will know that there have been many attempts at such reductive analyses of causation and none of these attempts has been completely successful. Every analysis faces cases where we intuitively judge that causation is clearly present or absent; yet the proposed analysis disagrees. Often, it is possible to modify the analysis slightly in order to correctly handle the counterexample, but, inevitably, new counterexamples will be found for the modified analysis. In the case of counterfactual analyses in particular, a small industry has developed searching for counterexamples and adding epicycle upon epicycle to the analysis in order cope with the counterexamples.

There are four possible responses to this unsatisfactory situation. First, we could simply continue on in the hope of discovering the perfect analysis which will agree with all our intuitive judgements and so have no counterexamples. The history of the field so far suggests that this hope will be in vain, but even if successful, the trajectory that analyses have taken to date would suggest that the successful analysis will be a rather complex affair with little appeal apart from the fact that it has been gerrymandered to agree with our intuitive judgements.

A second response is to take the failure of reductive analyses as evidence that causation is a conceptual primitive and cannot be reduced to non-causal concepts. There are two ways we could go here: we could take the relation c caused e (or perhaps cs cause es) as itself primitive (see Anscombe, 1971; Armstrong, 1997), or we could analyse this relation in terms of more basic—but nonetheless causal— notions. The problem with taking c caused e as a primitive relation is that such an 'analysis' is rather uninformative. In particular, it provides no guidance when answering the question of how it is that we recognize the causal relation in the world given Hume's ([1748], 1990, §VII) argument that we cannot directly perceive the obtaining of this relation. The alternative non-reductionist approach is to analyse the causal relation in terms of more basic causal concepts. Agency analyses, for example, are based on the idea that c caused e just in case bringing about c is an effective strategy for an agent to bring about e, where the notion of 'bringing about' is irreducibly causal (Menzies and Price, 1993). Non-reductive analyses of this stripe are sometimes criticized as being circular, and agency accounts have been said to be objectionably anthropocentric. I believe these charges can be adequately defeated (see, for example, Menzies and Price, 1993; Ahmed, 2007), but like reductive analyses, non-reductive analyses also face possible counterexamples (see, for example, Reutlinger, 2012), and so do not succeed at providing an analysis that agrees perfectly with our pre-theoretic intuitions.

A third response is to drop the requirement of perfect agreement with intuition. Rather than rating analyses on a pass/fail scale based on whether they agree with all our intuitions, perhaps we should judge them on a sliding scale based on how many of our intuitions they correctly capture. The motivating idea here is that conceptual analysis does not aim to simply describe our concepts: it aims to systematize them, to remove inconsistencies and bring them into reflective equilibrium. This view of conceptual analysis was explicitly endorsed by Lewis (1983b, x) and is probably the prevailing view amongst metaphysicians in the analytic tradition today. Despite being strongly guided by pre-theoretic intuition, then, conceptual analysis is seen as a potentially revisionist programme and so we should not be surprised, nor too bothered, if our analyses do not agree perfectly with intuition.

The problem with this third response is that we often face situations where two (or more) competing analyses seem to do equally well—each has its areas of discord with pre-theoretic intuition, but neither discord seems worse than the other. Lewis suggests that we should not assess each analysis individually, but should look at how it fits together with other conceptual analyses to form a coherent 'package deal'. Of course, this only defers the problem, for, as Lewis acknowledges, we may find that two (or more) competing package deals do equally well.

The three responses above share the assumption that there is a single coherent concept of causation that is the target of the conceptual analysis. The difficulties faced by each of these responses suggest a fourth response, which is to deny that there is a monolithic pre-theoretic concept of causation in the first place. This position is known as 'causal pluralism' and has been advocated by a small minority of philosophers (Sober, 1984; Skyrms, 1984; Hall, 2004; Cartwright, 2007; Godfrey-Smith, 2009; Longworth, 2010; Psillos, 2010). Let us look at this response in more detail.

8.1.1 Two Concepts of Cause

Perhaps the simplest and best-worked-out form of causal pluralism is that put forward by Ned Hall (2004), who argues that the word 'cause' is ambiguous: there is causation in the sense of *production*, and causation in the sense of *dependence*. Hall regarded dependence as being simply counterfactual dependence (if the cause had not happened, the effect would not have happened), but the cluster of intuitions identified with this concept can also be seen as motivating other reductive analyses of causation, with the relation of dependence being cashed out in other ways, for example as nomological dependence (cause and effect fall under a law); or as probabilistic dependence (the cause raises the probability of the effect). Perhaps for this reason, Peter Godfrey-Smith (2009) uses the term *difference-making* instead of dependence to refer to this sense of causation. Causation as production, on the other hand, is often left unanalysed, with the relation of production taken to be primitive (sometimes referred to as 'causal oomph'; see, for example, Kutach, 2014). Process theories of causation, as developed by Salmon (1997) and Dowe (1992; 2000), are an example of a non-primitive account of causation as production, analysing production in terms of the transfer of some physical quantity from cause to effect.

The two senses of causation are associated with separate groups of intuitions that are sometimes in tension with each other. Consider, for example, a situation in which Suzy and Billy each throw a stone at a glass bottle. Suzy's stone strikes the bottle, which then shatters. A split second later Billy's stone sails through the space the bottle used to occupy. Did Suzy's stone cause the bottle to shatter? Production intuitions seem to kick in here and suggest that the answer is yes. Suzy's stone produced, or brought about, the shattering, and this is true regardless of the fact that Suzy's stone did not make a difference to whether the bottle would shatter—if her stone had not been thrown, the bottle would have shattered anyway thanks to Billy's stone. Now consider a case in which Billy forgets to water his indoor plant, and it dies. Did the absence of water cause the plant's death? Dependence intuitions suggest that the answer is yes. Whether or not the plant has water makes a difference to whether or not it dies, and this is true despite the fact that an absence does not seem to be the kind of thing that can produce or bring about anything (for an absence is not a *thing* at all). In these two situations production and dependence criteria disagree on what caused what, suggesting that they cannot both be part of a single coherent concept of causation. Furthermore, common intuition seems to favour production in one case and dependence in the other, suggesting that they are, nonetheless, both attached to our ordinary use of the word 'cause'.

Hall points out that production and dependence do not only disagree on particular cases: they also disagree on some of the general properties of the causal relation. Causation as production is an intrinsic, local relation between cause and effect, whilst causation as dependence is an extrinsic relation. Further, causation as production is intuitively a transitive relation: if *a* brings about *b*, and *b* brings about *c*, then it seems natural to say that *a* has brought about *c* (albeit indirectly). Dependence, on the other hand, is not always transitive. Suppose, for example, that it is a hot day, and Tom and Winifred are having a water fight. Tom throws a water balloon at Winifred, who ducks and thereby avoids getting wet. In this situation we can suppose that it is true that Winifred's ducking depended—in the appropriate sense—on the water balloon being thrown (for example, we can suppose that she would not have ducked if the balloon had not been thrown). We can also suppose it is true that Winifred's remaining dry depended on her ducking (if she had not ducked, she would not have remained dry). But in most such situations it would be false that Winifred's remaining dry depended on the balloon being thrown (typical situations in which Tom does not throw his water balloon at Winnifred are situations in which Winnifred does not get wet).

Hall, then, has identified two inconsistent sets of intuitions surrounding causation. If our aim is to come up with a set of necessary and sufficient conditions for correctly stating that *c* caused *e*, and if we are not going to simply choose one set of intuitions and ignore the other, then the best we can hope for is a disjunctive analysis of causation (c caused e iff c produced e, or e depends in the right way on c). Some philosophers, however, have suggested that even a disjunctive analysis like this is too much to hope for. Nancy Cartwright (2007) argues that there are innumerable causal concepts, with each of our specific causal verbs (compress, attract, feed, ...) having its own distinctive causal content. Stathis Psillos (2010) argues that causation is like the common cold: we identify its existence by noticing various symptoms, and it is not necessary that all symptoms are present. Further, like the common cold, he argues, there is no deep underlying nature to causation that unites all these symptoms. Similar 'cluster concept' views have been advanced in passing by Brian Skyrms (1984) and Richard Healey (1994). Peter Godfrey-Smith (2009) argues that the defining characteristics of causation are permanently susceptible to being challenged and renegotiated. Christopher Hitchcock (2003) shows that even if we settle on a set of ingredients for a reductive analysis of causation (whether these ingredients be nomological dependence, probabilistic dependence, counterfactual dependence, or causal processes), there are many interesting relations that can be defined from these ingredients, each of which can appropriately be described as 'causal'. All of these philosophers agree that we will not be able to come up with an adequate analysis of causation in terms of a set of necessary and sufficient conditions for its application.

8.1.2   The Subjectivity of Causation

The considerations above suggests that conceptual analysis is not going to provide us with a single account of causation. Can empirical analysis do any better? Can we identify an objective feature of reality whose presence would make it true that c caused e? Given the inconsistent intuitions surrounding causation, it is hard to see how a single objective feature of reality could satisfy all of these intuitions. But perhaps this is not a bad thing; if there is an objective feature of the world that corresponds with one of the conceptual analyses above but no objective feature that corresponds with others, then this could break the tie between various conceptual analyses.

Unfortunately, as I will argue below, the correct application of the phrase 'c caused e' has an essentially subjective component. Whether it is correct to say that c caused e depends, in part, on the perspective and interests of the speaker, and this is a fact that cuts across the various groups of intuitions described above. This subjectivity in the application of the concept of causation suggests that we will not find an objective feature of the world that is picked out by the claim that c caused e.

There are actually a number of sources of subjectivity in our application of the concept of causation. I describe two of them below.

The Subjectivity of Selection

Much of the time we treat causation as a relation between an event and the (singular) occurrence that brought it about. Coroners try to determine the cause of death, technicians look for the cause of a malfunction, and we all might search out the cause of a delicious smell. There is some evidence that people from different cultures may disagree systematically in their judgements about what *counts* as a cause, but the existence of a concept of singular causation seems fairly universal across cultures (Choi et al., 1999; Bender and Beller, 2011; Le Guen et al., 2015). What is more, the idea of a single cause is put to work in scientific as well as everyday contexts. Consider, for example the following passage from psychologists Choi et al. describing what they call the Fundamental Attribution Error (FAE), a tendency to over-assign causal responsibility to dispositional traits of people and under-assign it to the context. They say 'the FAE may be said to occur when people infer a disposition corresponding to behavior under conditions in which *the true cause* lies in the situational context' (Choi et al., 1999, 46, my emphasis). In defining the FAE, Choi et al. rely on the idea of a singular cause bringing about an effect (and note that this is done while carefully considering mistakes that people intuitively make about causation).

A moment's reflection, however, reveals that the world is not as simple as this concept of causation makes out. An event is rarely, if ever, brought about by some single other event (unless that other event is an entire cross section of the universe at some earlier time—we will consider this thought below). The First World War, for example, was surely not the result of a single event (say, the assassination of Archduke Franz Ferdinand and his wife Sophie), but rather the coming together of numerous events: the assassination, various misunderstandings and miscommunications, and decades of political, territorial, and economic conflict between the European powers. Similarly, a person's behaviour is surely a result of *both* their inner dispositions and the context in which they find themselves, rather than one or the other. Even in a simple case such as the striking of a match, the striking is only part of what brings about the ignition of the match, for ignition also requires the presence of oxygen.

We could respond that our concept of causation allows for the possibility of multiple causes, and assert that 'the cause' when uttered in some context simply refers to the most salient cause in that context. But as Jonathan Schaffer points out, even when we allow multiple real causes we still typically draw a distinction between the real causes and mere background conditions:

> when the four movers collectively lug the piano up the stairs, it would be natural to select each individual mover's efforts as a real cause of the piano reaching the second floor (thereby selecting four real causes), all the while demoting various factors like the presence of the staircase to the status of being background conditions. (Schaffer, 2014, §2.3)

In applying the concept of causation, then, what we typically do is distinguish true causes from mere background conditions: the striking of a match is the cause of its ignition; the presence of oxygen is a mere background condition.

The problem is that the distinction between causes and conditions is not objective. It reflects the interests and context of the person making the judgement. As Herbert Hart and Tony Honoré point out, 'the cause of a great famine in India may be identified by an Indian peasant as the drought, but the World Food Authority may identify the Indian Government's failure to build up food reserves as the cause and the drought as a mere condition' (1959, 33).

In the words of John Stewart Mill:

> Nothing can better show the absence of any scientific ground for the distinction between the cause of a phenomenon and its conditions, than the capricious manner in which we select from among the conditions that which we choose to denominate the cause. However numerous the conditions may be, there is hardly any of them which may not, according to the purpose of our immediate discourse, obtain that nominal pre-eminence.   (Mill, 1843, V §3)

Although a few philosophers have argued that our inclination to distinguish causes from backgrounds is too predictable to be without a basis (Ducasse, 1926; Hart and Honoré, 1959), it is generally accepted that the distinction reflects our own interests rather than some distinction in the world. In his groundbreaking discussion of causation, for example, Lewis avoids discussing what he calls the 'invidious discrimination' between causes and conditions and focuses instead on 'the prior question of what it is to be one of the causes (unselectively speaking)' (1973a, 559).

But do we actually possess such a non-discriminatory concept of causation? Hart and Honoré (1959, 12), McDermott (2002, 94), and Schaffer (2005, 314) have argued that the selection of causes from background conditions is an essential part of our concept of causation. If they are right, and if the distinction between causes and conditions is interest-relative and does not reflect some objective feature of the world, then it follows that the concept (or concepts) of causation is essentially interest-relative.

The Problem of Non-Extensionality

A second sign of the non-objective, interest-relative nature of our causal intuitions is that whether we intuitively accept that one event causes another can depend on how those events are described. For example, we might in some situations be willing to assert that Billy's anger caused him to slam the door as he left the house, but deny that his anger caused him to close the door (he would have done that anyway). This is so even though 'Billy's slamming of the door' and 'Billy's closing of the door' pick out the same event in the actual world.

A second example, adapted from Yablo (1992): A pigeon has been trained to peck at all and only red objects. She is presented with a target of a particular shade of crimson, and she pecks at it. Intuitively, says Yablo, it is the target's being red that caused her to peck, not the target's being crimson. Being crimson is too specific to count as the cause.

A third example, from Achinstein (1975): it is intuitively correct that Socrates' *drinking hemlock* at dusk caused his death, but false that Socrates' drinking hemlock *at dusk* caused his death. Here, the difference in description is purely one of emphasis.

The fact that our intuitions are affected by the way in which events are described suggests that our intuitions are not latching onto an objective relation between events in the world. If this suggestion is correct, then an adequate conceptual analysis of *c caused e* will have to include the fact that whether or not *c* caused *e* is relative to our interests or descriptions or some such. Thus, rather than being a two-place relation between *c* and *e*, causation would turn out to be a three-place relation between *c*, *e*, and some subjective feature of the observer. Thus, if we are interested in the objective nature of the world, we should not look for objective relations of cause and effect, but rather look for objective features of the world that underlie our causal judgements.

8.1.3   Contrastive Causation

Jonathan Schaffer (2005) has recently advocated a theory of causation which he claims will solve the problems of selection and non-extensionality (among others). He claims that the problem is not that causation is a non-objective relation, but rather that we have misunderstood the nature and number of the relata. Rather than a two-place relation between events with the form '*c* caused *e*', Schaffer argues that causation is best understood as a four-place relation with the form '*c* rather than C^* caused *e* rather than E^*' where *c* and *e* are events and C^* and E^* are non-empty sets of contrast events. So, for example, we might say that smoking two packets a day rather than not smoking at all caused Jane to get lung cancer rather than have healthy lungs.

Schaffer's contrastive account of causation deals with the problem of selection in a straightforward way. Once a causal claim is spelled out in full detail, with the relevant contrasts made explicit, there simply is no further question of the distinction between causes and background conditions. The distinction is assimilated into the contrast classes. The problem of selection, therefore, only arises when a causal claim is made in the abbreviated form '*c* caused *e*'. Since the relevant contrast classes are not made explicit in such abbreviated claims, these claims are ambiguous. The correct disambiguation will be determined by context-sensitive pragmatic considerations: the claimant's intentions, interests, and beliefs, relevant social institutions, and the like.

The contrastive account of causation similarly relegates the problems of context sensitivity to pragmatics. When we assert a more abbreviated causal relation such as 'smoking two packets a day caused Jane to get lung cancer', the relevant contrast classes are left implicit and are determined by context. It is no wonder, then, that our intuitive judgements about abbreviated causal claims are also context dependent. Consider the Indian famine. In one context, when asking 'What caused the famine?', we may be interested in what caused the famine to occur now rather than some other time. In this context it would be appropriate to answer that the drought (rather than good rains) was the cause. In another context we may be interested in what caused the famine to occur in India rather than in some other similarly drought-affected nation. In this context it would be appropriate to answer that the government's failure to stockpile food (rather than stockpiling) was the cause.

The problem of non-extensionality is dealt with similarly. Different descriptions of the events mentioned in an abbreviated causal claim can suggest different relevant contrast classes. When we claim that Socrates' *drinking hemlock* at dusk caused his death, the emphasis suggests a contrast between drinking hemlock at dusk rather than doing something else at dusk. When we deny that Socrates' drinking hemlock *at dusk* caused his death, the emphasis suggests a contrast between drinking hemlock at dusk rather than drinking hemlock at some other time.

If Schaffer's suggestion is correct, then it follows that I am right that we will not find two-place cause-to-effect relations as part of the fundamental furniture of the world. But we should also be hesitant to admit four-place relations of contrastive causation as part of the fundamental structure of the world. According to Schaffer, causation relates events in the world with non-empty sets of contrast events. These contrast events are alternatives that could have taken place, but did not; they are mere possibilia. Unless we are willing to countenance merely possible events being part of the actual world, then, Schaffer's relation of contrastive causation cannot be part of the actual world.

Suppose, however, we are willing to admit that objective reality outstrips the actual, and in particular that merely possible events are part of objective reality even if they are not part of actuality. Then, relations of contrastive causation may be objective features of reality, but one then wonders, in virtue of what do these four-place relations of contrastive causation hold? The natural thought is that whether or not these relations hold will be determined by the properties and relations of concrete events in the actual world. Surely, the fact that dropping the vase rather than holding it causes the vase to shatter rather than stay in one piece is due entirely to the actual properties of the actual vase, its actual relation to the actual hard surface below it, the actual gravitational field, and so on. What is more, these same facts about the actual world will ground countless relations of contrastive causation. They also make it true, for example, that hurling the vase at the ground rather than squeezing it tightly will cause it to shatter rather than crack.

If causal statements are made true by a set of properties and relations in the actual world, and if, further, a single set of properties and relations in the actual world can account for the truth of a plethora of causal statements, then surely these properties and relations in the actual world should be the subject of our search for causation in the world. We might even justifiably call these properties and relations *causal* even if none of them qualifies as the relation of cause to effect. In the remainder of this chapter, and in the next, I will argue that our causal claims— ambiguous, interest-relative, or contrastive as the may be—are indeed grounded in a single set of properties and relations. In one way or another, true causal claims are all made true by the existence of appropriate causal influences and causal powers.

8.2 From Influence to Causation

So how exactly are causal influences related to causation? At first, it might be tempting to simply identify causal relations with the existence of causal influence, but such an attempt is doomed to failure. Causation is a relation, while causal influences are entities; relations are abstract, while influences are as real and concrete as tables, electrons, and electromagnetic fields.

If causation cannot simply be identified with influence, then we might be tempted to take the traditional approach and attempt to provide an analysis of causation in terms of causal influence. That is, we would look for an analysis that tells us that it is correct to assert that *c* caused *e* in a particular situation just in case that situation involves the existence of an appropriate structure of causal influence. However, I have just argued that there is no single concept of causation and, what is more, that our causal concepts do not pick out an objective feature of reality. Thus, we cannot hope to find an analysis of the causal relation in terms of causal influence.

My approach here is to argue for a weaker relation between causation and influence than identity or reduction. I shall do this in two steps. In the first step I will consider the cluster of intuitions surrounding the concept of causation as production. I will show that many of these intuitions can be satisfied by the presence of a certain structure of causal influence. Although this observation cannot be used to construct a set of necessary and sufficient conditions for the application of the concept of productive causation, we can use it to identify some rules of thumb for the application of the concept.

The second step considers causation as determination, and begins at the end of this chapter with a brief discussion of David Lewis's influential counterfactual analysis of causation. A full discussion of the idea of causation as determination takes place in the following chapter. My methodology there will be slightly different. In that chapter I will accept the increasingly popular view that causation as determination is best understood within the causal modelling framework as developed by Spirtes, Glymour, and Scheines (1993), Pearl (2000), and others. However, I will argue that the framework can, and should, be modified so as to explicitly represent causal powers and the influences that they manifest. In this way I hope to establish that, like our concepts of causation as production, concepts of causation as determination can, and should, be grounded in causal powers and causal influences.

8.3 Causation as Production

Intuitions surrounding causation as production, I believe, arise from situations in which there are just a few dominant influences, or a few particularly salient influences. To see this, let us focus first on an idealized situation in which there is only a single influence, I that is directed towards changes in some particular property of some basic object x. Now, as discussed in previous chapters, all causal influences are associated with one or more active powers. Basic influences are the manifestations of a single basic power; resultant influences are realized by a number of basic influences, and so will be the manifestation of a number of causal powers. Assume for the moment that I is a basic influence. Then, there will be a power P which belongs to part of some system (which includes x) such that the system being in state S triggers the causal power P to manifest the influence I. In this situation, I will argue, it makes perfect sense to regard S as the productive cause of the change in x (and so regard the change in x as the effect). Further-more, the influence I can be regarded as the physical embodiment of the causal relation. The reason that such an identification is justified is that I has all the hall-marks of productive causation. In particular, I will show that it has the following properties:

1. The influence 'connects' cause to effect, since the cause (which includes a causal power) is directed towards the existence of the influence, and the influence is directed towards the change in x.

2. The connection is a kind of necessitation, since the cause determines the effect through this influence (given our assumption that there is only one relevant influence).

3. The connection is asymmetric, since the effect does not determine the cause through this influence.

4. The connection is local, both in the sense that its existence is determined entirely by matters local to cause and effect, and in the sense that cause and effect typically involve only a few basic objects.

8.3.1   Connection

It is a common thought that causation serves as some kind of connection between cause and effect. As Hitchcock notes, 'There are many ways of attaching two objects together: for example, they can be connected, linked, tied or bound together; and the connection, link, tie or bind can be made of chain, rope, or cement. Every one of these means of attachment has been used as a metaphor for causation' (2003, 2). A metaphysics of power and influence allows us to step beyond mere metaphor and spell out the connection between cause and effect.

Consider, again, the idealized situation in which I is the only influence directed towards a change in some property of x. In this situation the connection between cause and effect is fairly straightforward. The system being in state S (the cause) triggers causal power P which is directed towards the manifestation of I. I, in turn, is directed towards a particular change in a particular property of a particular object of the system, namely x. Since I is the only relevant influence acting on that property of x, the change towards which I is directed will be the change that actually occurs (the effect). Thus, there is a connection of directedness, or, in George Molnar's (2003) words, *physical intentionality*, linking cause to effect.

In non-ideal situations, where I is only one of many influences directed towards changes in some property of x, the change towards which I is directed may not come about. However, I still forms a connection between the state S, and the change in x that actually occurs, since (as I argued in Chapter 6) I is literally a part of the total influence which *is* directed at the change that actually comes about.

Note that the connection I forms between cause and effect is not between *types* of cause or *types* of effect, but rather between the state of a *particular* system and changes in a *particular* object of that system. Thus, I is more closely related to singular causation than type-level causation.

8.3.2   Necessitation

The intuition behind the concept of causation as production does not simply require that there be some kind of connection between cause and effect: the connection must be one of production. The cause does not simply point to the effect: it brings it about. It is notoriously difficult to spell out what is meant by 'bringing about', and the relation is often taken as primitive (e.g. Anscombe, 1971; Tooley, 1987; Carroll, 1994; O'Connor, 2000), but it is also traditionally taken to imply some kind of necessitation; given that the cause occurred, the effect was determined to occur also (Spinoza, 1677, Part I, axiom 3; Kant, [1787] 1929, B5).

As with relations of productive causation, it is part of the concept of causal influence that they produce, or bring about, changes in the world (though, of course, they typically do this in combination with other influences). Influences are essentially directed towards a characteristic outcome, and they do not merely point towards the outcome the way an arrow might point to a location on a map. If an influence exists alone, it will bring about the outcome towards which it points. More generally, influences that do not exist in isolation contribute to producing the outcome that occurs. Causal influences, therefore, are characterized by the same idea of 'bringing about' that characterizes productive causation. It is hard to spell out in non-circular terms what is meant by 'bringing about' in this context, so, like the authors mentioned in the previous paragraph, I will take it as a primitive notion. However, once we put the primitive notion of bringing about in its proper place in the theory of causal influence, we can say a little more about it. In particular, the theory of causal influences tells us, contra Anscombe (1971), that there is a deep connection between *bringing about* and *necessitation*, and the theory allows us to spell this connection out.

To see the connection between causal influence and necessitation, it is, once again, easiest to begin with the idealized situation where there is only a single relevant influence. Basic powers, I have argued, are dispositions that are always successful; they cannot be finked and they have no antidotes. When triggered, they manifest in their characteristic way regardless of what else is going on. In our imagined situation, the system being in state S is the triggering condition for power P which is possessed by part of the system, and I is the characteristic manifestation of P. Thus, the system being in state S guarantees the existence of I. Furthermore, it is part of the essence of I that when it acts in isolation, the changes towards which it is directed will come about. Thus, the triggering of P—and hence the existence of I—brings with it a kind of conditional necessitation: given that I is the only influence of its type directed towards a change in some object x, the system being in state S will determine the change in x. A similar result follows if we allow probabilistic influence: given that I is the only influence of its type directed towards a change in object x, the system being in state S will determine a probability distribution over possible changes in the relevant property of x.

In the idealized situation, then, the connection of causal influence implies a conditional necessitation between cause and effect. But what of more realistic situations that involve more than one basic influence? In some situations it may be that one influence dominates over the others, in the sense that the outcome that occurs is similar to the characteristic outcome of the dominant influence, and variations (within some relevant range) in the other influences at play make little difference to the outcome. In such situations, we can, for many practical purposes, treat the dominant influence as if it were the only influence at play, and hence regard the dominant influence as conditionally necessitating the outcome.

In non-ideal situations where no single influence is dominant, things are not so different. Let B be the set of all basic influences of a given type that are directed towards a change in x at the time in question. Then, the members of B jointly constitute a total influence I on x, and this total influence conditionally necessitates the change that actually occurs in x. The necessitation here is still conditional, since it is possible that some relevant influence not in B could have been present, in which case I would not have been the total influence acting on x.[1] Thus, it is only given that the influences in B are the only relevant influences acting on x that their total I necessitates the change in x that actually occurs. Now, each influence in the set B is a part (in the sense discussed in Chapter 6) of this conditionally necessitating total influence I. Thus, we can say that each of these basic influences partially conditionally necessitates the outcome. The states of the various subsystems which trigger the basic powers that manifest the basic influences in B, therefore, also partially conditionally necessitate the outcome. In general, then, the connection provided by influences between various subsystems containing x and the change in x that occurs is a connection that implies a relation of partial conditional necessitation between the states of each of those subsystems and the outcome. This relation is not as strong as straightforward necessitation but it is surely all we could reasonably expect, given the obvious truth that causes rarely act in isolation.

8.3.3   Asymmetry

One of the most common intuitions about the relation of causation is that it is asymmetric: in most cases, at least, when c causes e, it is false that e causes c. Focusing on causation as production, these intuitions are captured by the thought that causes bring about their effects whilst effects do not bring about their causes. This suggests that the necessitation between cause and effect should also be asymmetric and this is indeed the case with the partial conditional necessitation provided by causal influence.

Suppose, again, that I is the total influence acting on x, and let S be the combined state of all the subsystems that manifest an influence that is a component of I. Then, S conditionally necessitates the change in x, and the states of the relevant subsystems partially conditionally necessitate this change. But the reverse is not true: the change in x does not conditionally necessitate the state S, nor does it partially conditionally necessitate the state of any of the subsystems.

To see this, note first that there *is* a reverse necessitation from the change in x to the total influence. Influences are characterized by the change towards which they are directed, and this change is guaranteed to occur if the influence acts in isolation. Thus, if we assume that the change in x was in fact produced through the action of causal influence, then the change in x determines the total influence that was acting on it—the total influence is guaranteed to be that which is directed

---

[1] Recall the discussion in Chapter 4 of Mumford and Anjum's (2011a) argument for dispositional modality.

towards the change that actually occurred. However, the reverse necessitation does not hold between the total influence and the states that manifest this influence.

There are two reasons for this failure of reverse necessitation. The first is that a resultant influence may be realized by many different sets of basic influences. Two partially trusted advisers together, for example, may have exactly the same influence as a single trusted adviser offering the same advice. Similarly, a force to the north combined with a force to the south could realize the same total as a force to the north-east combined with a force to the north-west. Thus, the total influence does not determine the basic influences which realize it, and so cannot necessitate the states of the subsystems that triggered the manifestation of these basic influences.

The second reason for the failure of reverse necessitation is that even basic influences do not determine the states of the systems that manifest them, since the same basic influence could be produced in different ways. A northerly directed force of exactly 1 N, for example, could be produced on an electron in any number of ways. A positive charge or a large mass appropriately placed to the north would do the job, as would a negative charge appropriately placed to the south. The presence of the force, therefore, does not necessitate any particular arrangement of particles. In general, knowing that a particular influence is present will not tell you the state of the system that manifested that influence (at least not without further assumptions about the range of possibilities).

8.3.4 Locality

Another characteristic intuition associated with the view of causation as production is that causation is local in two senses. First, causes and effects are relatively localized in time and space and involve a finite number of entities. So, for example, we identify Ali's punch as the cause of Foreman's fall, rather than citing the state of the entire stadium, or some even larger swathe of the universe, as the cause. The second sense in which causation as production is local (a sense disputed by counterfactual accounts) is that whether or not a caused b depends entirely on a, b and their relation to each other (and perhaps the chain of causes and effects linking a to b); it does not depend on anything extrinsic to a and b. So, this intuition tells us, whether or not Suzy's stone throw caused the bottle to shatter is a matter of the relationship between her throwing and the shattering (and possibly events in a chain linking the two) and has nothing to do with whether Billy was standing by ready to throw a stone, had Suzy missed.

Connections of causal influence are typically local in both these senses, for influences are manifested by causal powers, and, in line with Assumption 2 of Chapter 5, causal powers (or at least all causal powers that can be appealed to in a reductive explanation) must be of finite rank. Recall that a power has rank n

just in case the state which triggers the power is the state of a system of n basic objects. The powers that are referred to in reductive explanation typically have fairly low rank. Recall also that causal influences can be directed only towards changes in a single basic object. Thus, the connections of causal influence appealed to in reductive explanation can only hold between states of finite systems, and so connections of causal influence are local in the first sense. Further, remember that basic causal powers have no finks or antidotes. Whether and how a rank-n causal power manifests an influence is completely determined by the state of the appropriate n-object system; nothing extrinsic to this system makes a difference. So, connections of causal influence are also local in the second sense.

There is a technicality that I overlooked in the previous paragraph: basic objects are only basic with respect to a given reductive explanation. Thus, what is regarded as a basic object for the purposes of one explanation may turn out to be composed of an infinite number of parts with respect to some other explanation (or with respect to explanations at the fundamental level, if there is such a thing). Furthermore, nothing guarantees that basic objects will be localized in time or space (for example, we might take an infinitely extended electromagnetic field as a basic object in an explanation of cosmic evolution). Typically, however, the basic objects identified in a reductive explanation will be localized in time and space. Thus, reductive explanations typically posit connections of causal influence between localized systems that are, for the purposes of explanation, taken as finite. I believe this is enough to account for the intuition that causal connections are local.

8.3.5   Russell's Problem

It is worth noting that there is a tension between the intuition of necessitation and the intuition of locality. This tension was first noticed by Bertrand Russell (1913) and has been forcefully argued by Noa Latham (1987) and Hartry Field (2003). The problem is that if a cause is a localized event (or state), then it cannot necessitate its effect, since it is always possible that something outside the cause could intervene to stop the effect coming about. Flicking a switch causes a light to come on, but it does not necessitate that light coming on, since something (an errant bullet, say) could cut the circuit between the moment of the switch flicking and the moment of the light coming on. In order to rule out such outside interference, we would need to know the state of everything that could possibly affect the outcome, that is, we will need to know the state of an entire cross section of the backwards light cone of the effect. Thus, it is only entire cross sections of backwards light cones that can necessitate an effect. If we insist that causes necessitate their effects, then, we are forced to conclude that it is only entire cross sections of backwards light cones that can count as causes, and we violate the intuition of locality. If we insist on locality, then we must drop the claim that causes necessitate their effects.

The discussion above suggests that we take the second horn of this dilemma. Connections of causal influence are not connections of necessitation but connections of partial conditional necessitation. So, if, as I suggest, intuitions about causation as production are generated by connections of causal influence, then we might claim that causes only partially conditionally necessitate their effects. What is more, it is easy to see how an intuition involving a stronger form of necessitation can be produced by considering cases in which there is only a single dominant influence.

But Field (2003) sees a bigger worry here. Since an effect is necessitated only by an entire cross section of the past light cone, then *every* event in the past light cone must be counted as one of the causes of that effect. Sam's praying that the fire would go out, says Field, would be just as much a cause of the fire being extinguished as Sara's aiming the fire hose at it. Even worse, since there is always a possibility of the effect being prevented by some intervention from within the backwards light cone, the non-occurrence of those interventions must also be included among the causes of the effect. So Sam's not shooting a hole in the fire hose is also a cause, and, we might add, so is the fact that atoms in the surrounding environment did not spontaneously and improbably arrange themselves to form a Tyrannosaurus rex which then ate Sara. Surely, if we are to include all occurrences and all non-occurrences in the backwards light cone as causes, then the notion of causation loses its point.

I will not try to solve Field's problem here, since I am not trying to provide an analysis of causation. I will, however, point out that connections of causal influence do not suffer from the problems Field identifies: it is not the case that every event in the past light cone exerts an influence on a given effect. First, not all systems in the past light cone need have a causal power which does, or could, manifest an influence that is relevant to the effect. So it is false that all events (taken as the coming to be of states of systems) in the backwards light must be connected to the effect by causal influence. Second, causal influences have magnitudes, so there is a sense in which the state of one system can be more strongly connected to the effect than the state of another. Thus, even among those events that do exert an influence on some effect, it may make sense in some circumstances to pay attention only to those that exert the strongest influences. Finally, causal influences are real entities and are the manifestation of real powers. They are not manifested by non-occurrences, so it is false that non-occurrences exert any influence on the effect. In short, Field's arguments do not give us any reason to think that there is a problem with discussing links of causal influence. To the extent that causal talk is used to discuss connections of causal influence, then, causal talk will not lose its point in light of Field's worries.

8.3.6   An Analysis?

As Hitchcock (2003, 5–7) says, analysis of 'the causal relation' typically proceeds in two stages. In the first stage, a privileged class of entities is identified.
Laws of nature, relations of counterfactual dependence, probabilistic dependence,
manipulability, or transfers of conserved quantities are the usual suspects for
these privileged entities. In the second stage, 'c caused e' is analysed in terms
of the privileged class entities. I too have identified a privileged class of entities
(influences) and have argued that these entities underlie our concept of causation
as production. The close parallels between connections of causal production, on
the one hand, and connections of causal influence, on the other, justify my use of
causally laden language when discussing causal influence. For example, it justifies
my description of systems 'exerting' an influence, or influences 'acting on' an
object. So why not take the second step and try to come up with an analysis
of causation in terms of influence? I suggested above that there is more than
one concept of causation, but could we not focus on an analysis of causation as
production? Can we come up with an analysis of causal production in terms causal
influence? Given the close parallel between causation as production and influence,
we might try simply identifying causation as production with connection by causal
influence as follows:

(PC1) c is a productive cause of e just in case there is a connection of causal
influence between c and e.

Or better:

(PC2) c is a productive cause of e just in case there is a chain of connections of
causal influence between c and e.[2]

An analysis along these lines may be useful in some situations, and, I believe, it
will give the right answer in many cases that pose difficulties for other analyses.
It will, for example, accord with common intuition in many of the various cases
involving some kind of pre-emption. Inevitably, however, there will be situations
in which this analysis gives counter-intuitive results. Consider, again, the situation
discussed in Chapter 6 where Tybalt pushes Romeo towards a cliff edge whilst Juliet
pulls Romeo away from the cliff. Suppose that Tybalt is stronger and Romeo falls

---

[2] David Lewis's most recent attempt to define causation was stated in much this way, but he
meant something quite different by 'influence'. For Lewis, 'influence' denotes a complex relation of
counterfactual dependencies:

C influences E if and only if there is a substantial range C1, C2 ... of different not-
too-distant alterations of C (including the actual alteration of C) and there is a range
E1, E2 ... of alterations of E, at least some of which differ, such that if C1 had occurred,
E1 would have occurred, and if C2 had occurred, E2 would have occurred, and so on.

(Lewis, 2000, 190)

to his death. Clearly Tybalt's pushing is part of the cause of Romeo's death, and the proposed analysis above concurs: Tybalt exerted an influence that was directed towards a change in Romeo's motion, and the change in Romeo's motion resulted in his death. But the proposed analysis also counts Juliet's pulling as a cause of Romeo's death, since her pulling also exerted an influence directed towards a change in Romeo's motion, and the change in motion that resulted led to his death. Surely this conclusion is wrong: Romeo's death occurred *despite* Juliet's pull, not because of it.[3]

We could, of course, try to fix the analysis by further refining it. Taking into account the fact that 'cause' is often used as a success verb, we might analyse it as follows:

(PC3) c is a productive cause of e just in case there is a chain of connections of causal influence between c and e, and the influence that is constituted by this chain is directed towards e.

Here a chain of influences is taken to constitute a resultant influence, as described in Chapter 6.

The refined analysis correctly identifies Tybalt's push as a cause of Romeo's fall, while Juliet's pull is not, since Tybalt's push is directed towards an acceleration towards the cliff edge, while Juliet's pull is not. But this analysis will also face counterexamples. Consider, for example, a situation in which a barge is pulled eastwards by two horses, one on the northern bank of the canal, the other on the southern bank. The first horse will be exerting an influence that is directed towards an acceleration to the north-east. The second horse will be exerting an influence that is directed towards an acceleration to the south-east. The movement that actually occurs, however, is to the east. Thus, our refined analysis will pronounce that neither horse is a productive cause of the movement of the barge, a conclusion that seems clearly incorrect.

So, even if we try to restrict our attention to the concept of causation as production, I do not hold out high hopes for a successful analysis. Just as there are multiple concepts of causation, I believe that there are multiple concepts of causation as production. Each of (PC1), (PC2), and (PC3), for example, defines a perfectly good causal notion, and each of these might answer to the title 'productive cause' in a different situation. It is pointless, therefore, to look for an analysis of *the* concept of productive causation.

There is, however, something that (PC1), (PC2), and (PC3) all have in common: they are all concerned with the structure of connections of causal influence.

---

[3] This example also poses a problem for Lewis's analysis of causation in terms of influence. If Tybalt is only a little stronger than Juliet, then there are relevant alterations to Juliet's pull that would result in Romeo living. Hence Juliet's pull counts as an influence in Lewis's sense, and so is a cause of Romeo's death.

My claim is that it is this common concern for the structure of causal influence which makes each of (PC1), (PC2), and (PC3) *causal* concepts.

8.4 Causation as Dependence: The Counterfactual Analysis

The most well-known analysis of causation as dependence, and the most influential account of causation *simpliciter*, is David Lewis's counterfactual account (Lewis, 1973a). Lewis's analysis of causation proceeds in two steps. The first step is a definition of what Lewis calls *causal dependence*: let c and e be two events that actually occur, then e *depends causally* on c just in case e would not have occurred if c had not occurred. The second step defines causation in terms of causal dependence: let c and e be two events that actually occur, then c is a cause of e just in case either (i) e depends causally on c or (ii) there is a chain of causal dependence from e to c (that is, there are intermediate events $x_1, x_2, \ldots x_n$ such that x_1 depends causally on c, x_2 depends causally on x_1, $\ldots$ and e depends causally on x_n). So, for example, Tim throwing the rock caused the window to break, because the window would not have broken if it had not been hit by the rock, and it would not have been hit by the rock if Tim had not thrown the rock.

So far, there does not seem to be any role for causal influence in Lewis's analysis. In order to see that causal powers and influences do indeed play a fundamental role in the account, we need to delve a little deeper.

According to Lewis, causation is to be understood in terms of causal dependence, and causal dependence is just counterfactual dependence. That is, e depends causally on c just in case the following counterfactual conditional is true: e would not have occurred if c had not occurred. Since counterfactual conditionals are just as difficult to understand as causal relations, Lewis devotes much of his paper to developing an account of how to evaluate counterfactual conditionals like this. Put briefly, Lewis thinks that counterfactual conditionals should be understood in terms of similarity relations between possible worlds. His basic idea is that the proposition 'e would not have occurred if c had not occurred' is true just in case, among all the possible worlds in which c does not occur, the most similar worlds to ours are possible worlds in which e does not occur either (more precisely, some world in which c does not occur and e does not occur is more similar to the actual world than any world in which c does not occur but e does). The formation of certain footprints on the Moon at 2:56 UTC 21 July 1969, for example, depends counterfactually (and so causally) on the event of Neil Armstrong stepping onto the Moon at that time, since a world in which Armstrong did not so step on the Moon and no such footprints were formed is more similar to the actual world than any world in which Armstrong did not step on the Moon at that time but footprints were formed then anyway.

Thus, for Lewis, the truth of the claim that c caused e ultimately boils down to facts about relative similarity between possible worlds. So how, then, should we judge such similarity? What are the criteria? Lewis tells us that the relation of similarity is vague and suggests that it may be context-dependent, but he does give us some guidance. In judging overall similarity, he says, we must balance similarities in matters of particular fact against similarities of law. 'The prevailing laws of nature' he says 'are important to the character of a world; so similarities of law are weighty' (1973a, 560). But similarities of particular fact can also be weighty, he says: 'Comprehensive and exact similarities of particular fact throughout large spatiotemporal regions seem to have special weight. It may be worth a small miracle to prolong or expand a region of perfect match' (1973a, 560).

In general, says Lewis, the most similar worlds to ours will tend to be ones in which the laws and particular matters of fact are the same as ours up until very shortly before the putative cause c, at which point there is a small change (possibly a small 'miracle' that violates the laws), such that c does not occur. Afterwards, the laws of nature are once more fixed to match the actual world, and these laws determine what happens next in that world—possibly leading to large divergences from the actual world in terms of matters of fact.

It is here, in the consideration of the laws of nature across possible worlds, that the ontology of power and influence can enter into the counterfactual account of causation. For, if I am right that the laws of nature are grounded in an ontology of power and influence, then the requirement that we fix the laws of nature when evaluating counterfactuals (with the possible allowance of a small miracle), is a requirement that we fix the relevant causal powers and influences.

Of course, Lewis did not hold anything like my view of the laws of nature. Recall that Lewis defends a systems account of the laws of nature. According to the systems account, laws of nature are simply contingently true generalizations about the world. Of course not all true generalizations are laws ('all the coins in my pocket are worth 20 cents' might be a true generalization, but it is surely not a law of nature), so the systems account adds a further condition: 'A contingent generalization is a law of nature if and only if it appears as a theorem (or axiom) in each true deductive system that achieves a best combination of simplicity and strength' (Lewis, 1973b, 73). We need not go into the details of Lewis's best deductive systems. What is important here is that, for Lewis, the laws of nature are descriptions of the arrangement of particular local facts; there is no further fact that explains the regularities among these local facts or makes these regularities what they are. This feature of Lewis's account is known as *Humean supervenience*. The world, according to Lewis, ultimately consists in just a bunch of particular matters of fact—'one thing after another'; the laws of nature supervene on these facts, and do not add anything to them.

Now, Humean supervenience has been the target of many criticisms (for example, Tooley, 1977, 669; Carroll, 1994, 60–80; Maudlin, 2007, 67; Bird, 2007, 86).

One common concern is that the laws of nature are supposed to play an explanatory role, but, according to Humean supervenience, laws simply describe patterns among particular matters of fact and so are incapable of explaining these patterns. In the context of the counterfactual account of causation, this complaint gives rise to a concern that any account of laws based on Humean supervenience will leave it completely mysterious why, when analysing causal dependence, we are interested in what happens in possible worlds with laws similar to ours, rather than worlds with very different laws. The laws, according to the Humean, do not explain what gives rise to what (in any sense related to the notion of causation as production discussed above). The similarity of our world to those with the same laws consists only in similarities among the patterns of matters of fact, and we might wonder what, if anything, this can tell us about causation.

Things are quite different if the laws of nature are grounded in causal powers. For in that case the relation between the counterfactual account and causation as production is clear. If laws are grounded in an ontology of power and influence, then when we use the counterfactual account to answer whether c was a cause of e, we are effectively asking whether the influences at play would have brought about the effect if we removed the influence of c and replaced it with the most likely alternative. Thus, the counterfactual account, like the notion of causation as production, can be understood as concerned with structures of causal power and influence.

8.5 Conclusion

In this chapter I have argued that there is no single concept answering to 'the causal relation'. As Hitchcock (2003, 5) put it: ' "causation" is a subject matter, not a relation'. However, if causation is truly a subject matter, then there should be something that unites all the different concepts and discussions of causation. Following Hall (2004), I have gathered causal concepts into into two groups: causation as production and causation as dependence. I have then argued that the truth makers for causal claims from both of these groups are structures of influence. Thus, it is my contention that the subject of causation is defined by a concern with structures of causal influence.

My discussion of causation as dependence was quite brief, and I only focused on one conception of dependence: the counterfactual analysis of causation. One might wonder if the ontology of power and influence can play any role in other concepts of causation as dependence. Now, the concept of causation as dependence is usually analysed in terms of relations of counterfactual dependence or probabilistic dependence. In the past couple of decades, however, a different approach to causation as dependence has gained popularity. The idea is not to provide an analysis of the causal relation, but instead to develop 'causal models' which can be used to answer many different causal questions: Would e have occurred if c had not? Did c raise the probability of e? If so, by how much? What would have happened if d had occurred instead of c? What will happen if we do a? Was c a direct or indirect effect of e? Rather than a single analysis of the causal relation, causal models make precise a whole host of causal notions and provide techniques for answering questions about these notions. I believe that the best way to approach a more detailed investigation into the roles that power and influence can play in the concept of causation as dependence is via the causal modelling framework. That is the task of the next chapter.

9

Causal Models

The causal modelling framework aims to provide a set of tools for investigating the relations of causal dependence between events. Rather than providing a single analysis of causation as dependence, it provides tools for answering many different kinds of causal questions: Would *e* have occurred if *c* had not? Did *c* raise the probability of *e*? If so, by how much? What would have happened if *d* had occurred instead of *c*? What will happen if we do *a*? Was *c* a direct or indirect effect of *e*? And so on. We can, therefore, think of the causal modelling framework as a general approach to the concept of causation as dependence.

Causal models (like all models) must take some components as primitive. In particular, causal models rely on a primitive notion of 'causal dependence', and a causal model is essentially just a structure of such causal dependence relations. So what are these basic relations of causal dependence? Christopher Hitchcock (2001), James Woodward (2003), and Peter Menzies (2007) interpret them as relations of counterfactual dependence. If they are right, then we could regard causal models as providing analyses of causal concepts in terms of basic relations of counterfactual dependence. Judea Pearl, on the other hand, takes these relations of basic causal dependence to be primitively, and un-analysably, causal in nature. He says: 'I now take causal relationships to be the fundamental building blocks both of physical reality and of human understanding of that reality, and I regard probabilistic relationships as but the surface phenomena of the causal machinery that underlies and propels our understanding of the world' (Pearl, 2000, xv–xvi). If Pearl is right, then we can view the causal modelling framework as showing how various counterfactual and probabilistic dependencies can be grounded in fundamental causal relations.

In this chapter I will take Pearl's side of this disagreement, and argue that the fundamental causal relations required by causal models are, in fact, precisely the connections of causal influence that have been the topic of this book. In particular, I will argue that the standard causal modelling framework suffers from an important limitation, and I will present a modification to this standard framework that overcomes this limitation. In the modified framework, the basic relations explicitly represent basic causal powers and the influences that they exert. These 'causal influence models' can be used to generate standard causal models, and so can do everything that the standard causal models can do. I will argue, however, that there are both theoretical and practical reasons for preferring causal influence models over standard causal models. If I am correct, it follows that causal models can, and should, be seen as providing an understanding of the whole subject of causation as dependence in terms of basic causal powers and connections of basic causal influence.

9.1 Structural Causal Models

The causal modelling literature evolved out of attempts to answer causal questions in economics, epidemiology, and other social sciences, where the aim was to find a systematic mathematical way to pose and answer causal questions. Now, a central idea of the causal modelling literature is that in order to answer causal questions, we must first have an explicitly causal model of the situation—mere statements of observed association will not do. Thus, it is necessary to develop a formal language to describe, and pose questions to, these causal models. An early development of the method can be found in Spirtes et al. (1993), though the state-of-the-art presentation of this approach is Judea Pearl's *Causality* (2000), in which he presents what he calls the Structural Causal Model (SCM) framework for dealing with causation. The causal modelling framework has revolutionized statistical approaches to answering causal questions in many fields. It has also become popular amongst philosophers as a tool for understanding the nature of causation (see, for example, Hitchcock, 2001; Woodward, 2003; and Menzies, 2007). In this section I will give a brief overview of the SCM framework.

Formally, the SCM framework consists of two parts: a set of variables and a set of equations. The variables are typically represented by italicized upper case letters, $A, B, C, \ldots$, and represent relevant properties of the system that is being modelled. In the simplest case, these variables will take the value 0 or 1. So, for example, if we were studying the causes of lung cancer, we might use the variable C to represent the presence or absence of cancer in an individual, with the value 1 representing presence and 0 representing absence. In more complex situations, variables might take values from a larger discrete set or even a continuous range of values.

The second element of a causal model is a set of so-called *structural equations* which represent causal dependencies between variables. There are the same number of equations as there are variables, and each equation gives the value of one of the variables, which appears alone on the left-hand side of the equation. So each variable appears on the left-hand side of one, and only one, equation. If a variable does not have any causal dependencies on other variables in the model, it is said to be an *exogenous* variable, and the structural equation for that variable will simply state the actual value of that variable, e.g. $X = 1$. If a variable does have some causal dependence on other variables in the model, it is said to be an *endogenous* variable, and its structural equation will take the form:

$$Z = f_z(X_1, \ldots, X_n) \qquad (9.1)$$

where $X_1, \ldots, X_n$ are all and only the variables that play a role in causally determining the value of the endogenous variable Z.

Structural equations cannot be treated as normal algebraic equations, because, unlike algebraic equations, structural equations have a direction. They tell us that the variable on the left is causally determined by the variables on the right; they do not imply determination in the reverse direction. So, for example, suppose that X stands for the presence of a particular virus, Y stands for the presence of a typical symptom of the virus, and U_Y stands for any other factors that might possibly affect Y. Then we might have:

$$Y = \beta X + U_Y \qquad (9.2)$$

for some constant β. We cannot rearrange this equation to give:

$$X = (Y - U_Y)/\beta \qquad (9.3)$$

since this would incorrectly imply that the presence of the virus is partially determined by the presence of the symptom. Pearl interprets the directionality of the structural equations as representing the directionality of causation: causes bring about their effects, not vice versa. Woodward and Hitchcock, on the other hand, see structural equations as representing counterfactual dependencies, and the directionality of the equations represents the fact that (in most situations, at least) we rule out backtracking counterfactuals: we are happy to say that if the patient had not had the virus, she would not have had the symptoms; but we are less happy to assert that if she had not had the symptoms then she would not have had the virus. I will return to the difference between these two interpretations later.

The structure of causal determination that is encoded in the structural equations of a model can also be represented graphically. First, draw a node corresponding to each variable in the model. Then go through the equations and draw an arrow from the node representing each variable on the right-hand side of the equation to the variable on the left. Thus, an arrow from X to Y represents the fact that at least some values of X can play a role in determining the value of Y (and more importantly when it comes to using causal models to do statistics, the absence of an arrow from X to Y represents that X plays no role in determining Y).

Consider, for example, the causal relationship between local atmospheric pressure and the reading on a barometer. Let A represent the atmospheric pressure, and B represent the reading on the barometer, then we can represent the causal relation between them via the following structural equation:

$$B = A \qquad (9.4)$$

To get a complete causal model of this simple system, we also need a structural equation for A. If we take the atmospheric pressure to be an exogenous variable,

Figure 9.1. A simple causal model

then its structural equation will simply set its value to the actual value, say:

$$A = 900\,\text{hPa} \tag{9.5}$$

To make things more interesting, let us add a third variable, S, which represents whether or not a storm occurs ($S = 0$ represents no storm, $S = 1$ represents a storm), and make the unrealistic assumption that a storm will definitely occur if the pressure drops below 1000 hPa. This relation can be represented by the following equation:

$$S = \begin{cases} 1 & \text{if } A < 1000\,\text{hPa} \\ 0 & \text{if } A \geqslant 1000\,\text{hPa} \end{cases} \tag{9.6}$$

The graph for this causal model is shown in Figure 9.1. In this graph, The arrow from A to B represents the causal dependence of B upon A stated in the equation 9.4, the arrow from A to S represents the causal dependence of S upon A stated in equation 9.6, and the fact that there are no arrows pointing towards A represents that A is an exogenous variable, as stated in equation 9.5.

9.1.1 Answering Questions about Counterfactual Dependence

Once we have a causal model, we can use it to make predictions and to answer questions about what would have happened had things been different. The key to doing this is the notion of an *intervention*, which is a surgical change to the value of one or more of the variables. To model an intervention on a variable, we simply replace the structural equation for that variable with an equation that assigns the new value. All other equations are left unchanged. For example, if we want to know what would have happened if the atmospheric pressure had been 1100 hPa (or what will happen if the pressure reaches that level, whilst everything else remains the same), then we simply replace the structural equation $A = 900$ hPa with $A = 1100$ hPa. We then solve the remaining equations to give $B = 1100$, $S = 0$. We interpret this as telling us that if the atmospheric pressure had been 1100 hPa, then the barometer would have read 1100 and there would have been no storm.

If, instead, we want to know what would have happened if the barometer had read 1100, then we replace the equation $B = A$ with $B = 1100$. Solving the remaining equations gives $A = 900$ and $S = 1$. So, given that the actual air pressure is 900 hPa, if the barometer had read 1100, then the atmospheric pressure would still have been 900 hPa and there would still have been a storm (that is, if the barometer had read 1100, it would have been wrong).

A causal model together with the concept of an intervention provides us with a straightforward and principled way to evaluate counterfactuals. So, if the relation of cause and effect is built out of relations of counterfactual dependence, then a causal model provides us with the information we need in order to figure out what causes what. For example, if causation simply is counterfactual dependence, then our model tells us that atmospheric pressure is a cause of the storm, but the barometer reading is not, since an intervention on A can make a difference to S, whilst an intervention on B cannot.

9.1.2   Answering Questions about Probabilistic Dependence

So far, we have been assuming that structural equations are deterministic. If we also assume that the model is acyclic (which is to say that if you start at any node in the graph and trace out a path by following arrows from node to node in the direction that they point, you will never end up back where you started), then there is always a unique solution to the values of all the endogenous variables once given the values of all the exogenous variables. In a deterministic acyclic model, then, the values of the endogenous variables are uniquely determined by the values of the exogenous variables. In most real-world applications of causal modelling, however, we do not have complete knowledge of the values of the exogenous variables. In such cases, we cannot write down structural equations that set all the exogenous variables to their actual value; the best we can do is specify a probability distribution over the possible values of each exogenous variable. But given a probability distribution for each exogenous variable, a deterministic acyclic causal model will determine a unique probability distribution for each endogenous variable. It is for this reason that techniques involving causal models can be very useful in studying statistics. Causal models can be used to estimate the probability of a desired outcome given various possible interventions, to estimate the probability that an observed effect would have occurred if we had done things differently, and so on.

The probabilities mentioned above are all epistemic; they result from our lack of complete knowledge regarding the values of the variables. But structural causal models can accommodate genuine indeterministic relations as well. For example, stimulated emission is a process in which the presence of an electromagnetic field at a certain frequency causes an electron to jump from a high-energy state to a lower-energy state, emitting a photon in the process. The equations of quantum mechanics assign non-trivial probabilities to such a transition given the presence of the stimulating electromagnetic field. An obvious way to interpret this fact is to regard the causal interaction between the electromagnetic field and the transition of the electron as irreducibly probabilistic. Truly 'chancy' probabilistic causal interactions of this kind could be modelled in the SCM framework by allowing probabilistic structural equations. So, for example, if A is the amplitude of the electromagnetic field at the relevant frequency, and T is a binary variable representing whether or not the electron transition occurs ($T = 1$ indicating the transition occurs), then the endogenous structural equation for T could be written as:

$$\mathrm{Prob}(T = 1) = bA \qquad (9.7)$$

(The constant b in this case is known as the 'Einstein B Coefficient'.)

Equivalently, we can model this probabilistic relation without modifying the SCM framework by introducing a new exogenous variable U, whose value is unknown, but is assigned a probability distribution. So, for example, if U is a variable which has probability b of taking the value 1, and takes the value 0 otherwise, then we can replace equation 9.7 with the following deterministic equation:

$$T = UA \qquad (9.8)$$

The probability distribution for U will then determine a probability distribution for T which agrees with equation 9.7.

Thus, the SCM framework can be used to investigate relations of probabilistic dependence regardless of whether the probabilities involved are epistemic or ontic.

9.2 Level Invariance and Modularity

Whether we are interested in relations of counterfactual or probabilistic dependence, the SCM framework makes heavy use of intervention. It is the ability to ask and answer questions about what would happen under certain interventions which separates causal models from merely statistical models. However, the way that interventions are represented and used in the SCM framework relies on a crucial assumption, namely that the model will continue to apply otherwise unaltered when an intervention takes place. Being a bit more precise, this assumption can be separated into two independent assumptions, one about the individual equations in the model, the other about the system of equations as a whole.

The first of these assumptions is that each equation satisfies a condition that Woodward (2003, 322) refers to as *level invariance*, defined as follows:

Definition 16 (Level invariance). An equation is *level invariant* if it (or the relationship that it represents) continues to hold under some relevant range of interventions which set the values of some of the variables on the right-hand side.

Suppose, for example, that we have observed that the height to which a certain kind of plant grows is proportional to the amount of water it receives each week. This relationship of proportionality would be level-invariant if it continues to hold in situations in which we intervene to set the amount of water the plant gets rather than simply observing. If this condition was not satisfied, then clearly we could not use the equation describing this relationship to draw conclusions about what would happen under an intervention. As Woodward points out, however, we do not need the relationship to hold under all possible interventions; it need only hold under a relevant range of interventions. Any relationship of proportionality between plant height and water received, for example, would surely break down in situations where the plant was given hundreds or thousands of litres of water each week. Nonetheless, if the relationship holds for interventions in the range between 1 and 3 litres per week, then we can use the equation to tell us what would happen under an intervention within this range.

The second assumption that is required in order to use structural equations to evaluate counterfactuals is that the system of equations is *modular* in the following sense:

Definition 17 (Modularity). A system of equations is modular just in case, for each equation in the system, there is a possible intervention which changes that equation (by setting the dependent variable) without thereby disturbing any of the other equations in the system.[1]

If a system of equations is not modular, then we cannot use the method described above to evaluate counterfactuals, since there will be no way to intervene to set one of the variables without changing relationships between the other variables.

Do we have any independent reason for believing that causal models will be level-invariant and modular? That is to say, is there anything about the relations that are represented in a causal model which would justify the claim that they can only be correctly represented by a causal model that is level-invariant and modular? Woodward thinks so. It is a natural thought, he says, that if a system of equations 'correctly and fully represents the causal structure of some system', then each equation in the system 'should represent the operation of a distinct causal mechanism' (2003, 48). Furthermore, he adds, it is plausible to insist that two mechanisms are distinct only if it is possible, at least in principle, to interfere with the operation of one without interfering with the operation of the other, and vice versa. Given these two assumptions, it follows that any causal model which correctly and fully represents the causal structure of a system will be level-invariant and modular.

---

[1] Pearl (2000) uses the term 'structural' to describe a system of equations satisfying this condition. Woodward's (2003, 329) definition of modularity is the same as mine except that he adds the condition that each equation in the system be level-invariant.

9.3 Mechanisms and Influence

So, according to Woodward, it is the fact that distinct equations represent distinct causal mechanisms that justifies our belief that a correct causal model will be both level-invariant and modular. But what exactly is a causal mechanism, and why is it plausible to think that it should be possible to interfere with one mechanism without interfering with another? Woodward does not give a definition of 'causal mechanism'; he relies instead on existing intuitions regarding the concept along with various examples to motivate the claim that distinct causal mechanisms will be independent in the sense mentioned above. I have no real complaint against this approach. After all, every argument will have to fall back on a common intuitive understanding of *something*. However, I do think that the SCM framework tends to leave out important information about the causal mechanisms (intuitively understood) that are at play in the systems modelled. Paying attention to this information, I argue, will give us a better understanding of causal mechanisms and why they are independent, and will suggest a modification to the SCM framework that will make the framework more useful.

According to Woodward, each structural equation (or, equivalently, each complete set of arrows directed into a variable in a directed graph) should represent a distinct causal mechanism. There are many situations, however, in which a single structural equation will (intuitively) represent more than one causal mechanism. Consider a situation in which Annabel and Billy are independently considering baking birthday cakes for Charles, and suppose that nobody else will bake him the cake. Then, the number of cakes that Charles receives for his birthday will be determined by the number that Annabel and Billy bake. We can capture this situation with a causal model as follows. Let A represent the number of cakes that Annabel bakes, let B represent the number of cakes that Billy bakes, and let C represent the number of cakes that Charles receives. Then the structural equations for this situation will be:

$$A = a \tag{9.9}$$

$$B = b \tag{9.10}$$

$$C = A + B \tag{9.11}$$

where a and b are the values A and B take in the situation being modelled. The graph for this situation is given in Figure 9.2.

According to Woodward, equation 9.11 represents a causal mechanism. I believe, however, that intuition dictates that this equation represents the action of *two* causal mechanisms: one that operates through Annabel, and one that operates through Billy. What is more, these mechanisms are distinct in the sense identified by Woodward. It is possible to interfere with Annabel's cake baking (say by taking

Figure 9.2. Charles's birthday cake

her out to the cinema when she had planned to bake the cake) without interfering with Billy's cake baking, and vice versa.

In general, I suggest, a structural equation does not represent a single distinct causal mechanism. Rather, it represents the outcome of all the distinct causal mechanisms that act upon a single variable. To be clear, my point is not that Woodward is wrong in his claim that there is a distinct causal mechanism represented by equation 9.11, for it is possible that a causal mechanism could be composed of distinct sub-mechanisms (just as influences can be composed of more basic influences: see Chapter 6). Rather, my point is that the system of equations seems to leave out important information about what we might call the *basic* mechanisms involved in the model (I will show why this information is important below).

Woodward is not ignorant of the fact that a single structural equation may represent more than one distinct mechanism. Hausman and Woodward (1999, 547) note that if a structural equation is linear, or more generally 'additive', then it will be possible to interfere with any one of the coefficients in the equation without affecting the way the other coefficients affect the outcome. This means that each coefficient in the equation will represent a distinct mechanism. Equation 9.11 is an example of such an 'additive' equation. The addition operation in this equation means that we can remove B from the equation without affecting the role that A plays in the equation, and vice versa. In graph theoretical terms, breaking the arrow from B to C leaves the arrow from A to C undisturbed and vice versa. In general, if the structural equation for a variable is additive, then every arrow directed at that variable will represent a distinct causal mechanism. Breaking any one of these arrows will leave the other arrows undisturbed.

If the structural equations for a model are all additive, then every arrow in the graph of the model represents a distinct causal mechanism and these mechanisms will be basic in the sense that if any one of these mechanisms is composed of sub-mechanisms, these sub mechanisms cannot be represented without adding additional variables to the model. If the equations are not additive, it may be the case that each equation represents a mechanism that is not composed of more basic mechanisms. In such a case (as with the additive case) the causal model will adequately represent all of the basic mechanisms involved. In many (and, perhaps, most) cases, however, the basic mechanisms will lie at a level between that represented by the arrows, on the one hand, and the structural equations, on the other.

Consider, for example, a situation in which three charged particles are interacting in the absence of any other significant influences (in particular, we will assume that gravitational interactions are negligible). For simplicity, we will assume that the particles are located in a line, with particle 1 on the far left and particle 3 on the far right. Let Q_1, Q_2, and Q_3 represent the charges of the three particles, D_1 and D_2 represent the distance between particles 1 and 3, and 2 and 3 respectfully, M_3 represent the mass of the third particle, and finally let A_3 represent the instantaneous acceleration of particle 3 at the time being considered (where a positive value indicates acceleration to the right). Coulomb's law, together with Newtonian mechanics tells us that the structural equation for A_3 is:

$$A_3 = \frac{k_e}{M_3} \left(\frac{Q_1 Q_3}{D_1^2} + \frac{Q_2 Q_3}{D_2^2} \right) \tag{9.12}$$

where k_e is Coulomb's constant. The graph for the causal model of this situation is shown in Figure 9.3.

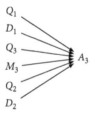

Figure 9.3. SCM graph of three charged particles

It can be seen that there are six arrows leading in to A_3. These arrows correctly represent that, in the situation modelled, there are six ways to manipulate the value of A_3. Now, equation 9.12 is not additive (since it involves products and quotients of variables), so we cannot regard these arrows as representing six independent mechanisms. Nonetheless, there does seem to be more than one basic mechanism involved. Intuitively there are *two* basic causal mechanisms. The electromagnetic force resulting from the interaction of particles 1 and 3 is one mechanism, and the electromagnetic force resulting from the interaction of particles 2 and 3 is another mechanism. These mechanisms are distinct, since you can interfere with one (say, by removing the first particle) without interfering with the other. The relations represented by the six arrows, on the other hand, are not distinct mechanisms. If you break the arrow between Q_1 and A_3, for example (say, by neutralizing the charge of the first particle), then you will also break the arrow from D_1 to A_3, since a neutral particle will not exert an electromagnetic force on particle 3 regardless of the distance between them.

Thus, there is a structure of distinct basic causal mechanisms that, in general, lies at a level somewhere between the arrows found in causal graphs and the structural equations that summarize the action of all the incoming arrows on a variable. Furthermore, the last example is rather suggestive. The basic mechanisms in this example involved the action of forces, and forces are paradigm examples of causal influence. It is my contention that, in general, the basic causal mechanisms that have been the topic of this section are nothing other than the manifestation of causal influence by causal powers. Recall from Chapter 4 that the influence which a causal power exerts depends only upon the variables within its domain. So it will be possible to interfere with the action of one causal power whilst leaving the action of all other causal powers untouched, so long as your intervention does not affect any variables in the domain of these other causal powers. For this reason the manifestation of an influence by a causal power seems to be the right kind of thing to count as an independent causal mechanism.

These thoughts motivate a modification of the standard SCM framework which I will call the *Causal Influence Model* (CIM) framework. This modified framework introduces a graphical method for keeping track of the basic mechanisms involved in a model and interprets these basic mechanisms as the manifestation of influence by basic causal powers. I will present this framework in the next section and then argue that it has several advantages over the standard SCM framework.

9.4 Causal Influence Models

Causal mechanisms, I claim, correspond to causal influences. Each distinct basic mechanism for changing a variable corresponds to a basic causal power that exerts an influence on that variable. As discussed in Chapters 4 and 5, the influence that a causal power exerts will depend on the state of a subsystem of the total system in which the power sits. If we represent the state of the total system with a set of variables, one variable for each relevant property of the system, then the state of the subsystem that determines the influence exerted by a given power can be represented by one or more of the variables in this complete set. So, for example, the electromagnetic influence between two particles is responsive to the charge of each particle and the distance between them, but not to their colour. In general, then, causal powers connect the values of multiple variables to influences directed at changes in the value of a single variable. This kind of structure can be represented graphically using an arrow that has a single head, but multiple tails. The tails will come from the variables that represent the state of the relevant subsystem; the head will point at the variable that is being influenced.

If we modify the standard SCM graphs to allow arrows with multiple tails, then we can insist that each arrow represent a distinct basic mechanism (that is, the manifestation of an influence by a basic causal power). Let us call such a graph a *causal influence graph.* The causal influence graph for the system of three charged particles described above is shown in Figure 9.4. This graph makes it very clear that there are two separate basic mechanisms involved, though each mechanism could be manipulated by intervening on any of four different variables. This example also shows that some variables, like Q_3, can be involved in more than one mechanism.

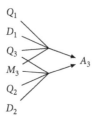

Figure 9.4. Causal influence graph of three charged particles

A causal influence graph can easily be turned into a standard causal graph by replacing each multi-tailed arrow with a multitude of single-tailed arrows. We simply draw a single-tailed arrow coming from each variable that lies at a tail of the multi-tailed arrow, and pointing to the variable which lies at the head of the multi-tailed arrow. If, after doing this for all multi-tailed arrows, we have more than one single-tailed arrow connecting two variables in the same direction, then we simply delete all but one of these arrows. Applying the procedure to Figure 9.4, for example, gives back Figure 9.3.

With this procedure, every causal influence graph determines a unique causal graph of the standard SCM kind. What is more, a causal influence graph and the standard graph that it determines will agree on which variables can be intervened upon to manipulate the value of any given variable. Thus, the procedure preserves information about the counterfactual dependencies between variables. On the other hand, a standard causal model does not uniquely determine a corresponding causal influence model. Two or more distinct causal influence graphs may be associated with a single standard causal graph. For example, the causal influence graphs in Figures 9.5a and 9.5b are both associated with the same standard causal graph, which is shown in Figure 9.5c.

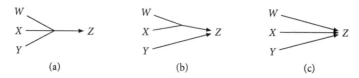

Figure 9.5. Distinct causal influence graphs that correspond to the same SCM graph

Causal influence graphs, therefore, contain all the information that is contained in standard causal graphs plus a little bit more. Both kinds of graph contain information about which variables depend counterfactually upon which, but only causal influence graphs contain information about the distinct basic mechanisms involved.

The new structure represented by causal influence graphs will have consequences for the structural equations that represent them. I will discuss these consequences below. In the meantime we should note that the structural equation for a given variable represents the combined effect of all the basic causal influences acting on that variable in a given situation. To put this another way, the structural equation for a variable represents the total influence manifested by a compound causal power whose components are all the basic powers that exert influence on the property represented by the variable.

Causal influence graphs, I believe, capture the intuitive concept of a basic causal mechanism that is left out of of standard SCM graphs, and I have argued that these causal mechanisms can be identified with the action of causal powers that are the topic of this book. But is there any benefit for the causal modelling framework to be gained by capturing this information? Given the important role that causal influences play in the reductive method of explanation, we might expect that there is indeed a benefit, but, on the other hand, the structural causal model framework seems to have done very well whilst completely ignoring causal influences. So what, if anything, do we gain by paying attention to basic causal mechanisms of the sort suggested here?

9.5 Grounding Counterfactuals

The first advantage we gain by paying attention to causal influence is that we get a better understanding of the relations of counterfactual dependence that are the subject of the standard SCM framework. As Woodward and Hitchcock point out, structural equations are not simply mathematical formulae: they are to be interpreted 'physically' as describing counterfactual dependencies between variables:

> They do not just represent patterns of association: instead, each deterministic equation is understood as encoding counterfactual information about how Y would change under manipulations of its direct causes.   (Woodward, 2003, 43)

This physical interpretation of the equations explains why they are to be treated asymmetrically. As Hitchcock puts it:

> The syntax of such an equation is richer than that of ordinary mathematical equations. In particular, structural equations are not symmetric: $Z = f_Z(X, Y, \ldots, W)$ is not equivalent to $f_Z(X, Y, \ldots, W) = Z$. (In structural equations, side matters.) This is because structural equations encode counterfactuals. For example, $Z = f_Z(X, Y, \ldots, W)$ encodes a set of counterfactuals of the following form:

If it were the case that $X = x$, $Y =,\ldots,$ $W = w$, then it would be the case that $Z = f_z(x, y,\ldots, w)$. These counterfactuals are to be understood along the lines discussed in section I above; in particular, they do not backtrack.   (Hitchcock, 2001, 280)

Note, though, that a system of structural equations does not directly represent *all* counterfactuals that are true of the system. Consider, for example, the following system:

$$A = a \tag{9.13}$$

$$B = A \tag{9.14}$$

$$C = B \tag{9.15}$$

Here, equation 9.14 tells us what would happen to B if we were to manipulate A, and equation 9.15 tells us what would happen to C if we were to manipulate B, but none of the equations tells us directly what would happen to C if we were to manipulate A. On the other hand, equations 9.14 and 9.15 together give us the resources to *calculate* what would happen to C if we were to manipulate A. In general, Hitchcock tells us, the set of structural equations for a system 'is a set of *fundamental* equations from which all other counterfactuals may be derived' (2001, 283).

Hitchcock's assertion raises two questions: What, if anything, is the significance of saying that these equations are fundamental? And why should we think that the structural equations will allow us to calculate *all* counterfactuals that are true of the system?

In answer to the first of these questions, Hitchcock and Woodward tell us that the fundamental equations are those that are level-invariant and modular. I will return to this point below. In the meantime, note that there is a simple, intuitive answer to both these questions if we understand the structural equations for a system as representing all and only the basic causal powers at play within that system (or, at least, all the significant powers). The fundamentality of the equations is due to the fact that they capture the combined action of only *basic* powers, and the comprehensiveness of the equations is due to the fact that they capture *all* significant basic powers. Recall that Assumption 3 of the reductive method states that the behaviour of any object in a system is completely determined by the set of all the basic influences acting upon it (or, in the non-deterministic case, the probability distribution of behaviours is so determined). If this assumption holds, then all questions about the behaviour of a given object can be answered by keeping track of all the basic causal influences acting on that object. Since a structural equation keeps track of all the influences acting on a given variable, the set of structural equations for a system can be used to answer all questions, including counterfactual questions, about the behaviour of the variables represented, so long as all the relevant and significant variables are included in the model.

In fact, we can take this point further. Structural equations encode counterfactual dependencies, and it is because these counterfactual dependencies do not backtrack that the equations cannot be treated symmetrically. The SCM framework does not tell us why these counterfactuals hold, nor does it tell us why they must be non-backtracking counterfactuals. Now, adherents to Lewis-style counterfactual analyses of causation may not be too concerned by this lack of explanation, since they take counterfactual dependencies to be essentially primitive and causation is then simply *defined* as typically involving non-backtracking counterfactuals.[2] For those of us who are unsatisfied with the non-answers to these questions provided by counterfactual analyses, the causal influence framework can help.

First, by keeping track of causal influences, the causal influence framework identifies the physical mechanisms that underlie counterfactual dependencies. The graph in Figure 9.4, for example, shows that the counterfactual dependence of A_3 on the other variables is mediated through two causal influences. Each of these influences involves a number of variables, and this group of variables identifies the physical subsystem that has given rise to the influence. So, for example, Figure 9.4 shows that the counterfactual dependence of A_3 on D_1 is due to an influence that arises through the interaction of the charges of particles 1 and 3, the distance between the two particles, and the mass of particle 3. This information already provides some understanding of the origin of the counterfactual dependence of A_3 on D_1. We get an even more complete explanation if we accept the ontology of causal powers suggested in this book. Causal powers are essentially dispositional, and essentially dispositional powers serve as the truth makers for the counterfactuals that describe them: the fact that two positively charged particles would repel each other if placed in close proximity is due to the essentially dispositional nature of electric charge. By identifying the causal powers that interact to produce each influence, causal influence models show how counterfactual dependencies can be traced back to the essentially dispositional nature of these powers.

Tracing counterfactual dependencies back to the influences that are responsible for them also provides a partial explanation of why it is that the counterfactual dependencies in causal models are non-backtracking. The explanation comes down to the fact that causal influences are essentially asymmetric. As argued in the previous chapter, the (partial, conditional) necessitation produced by causal influence flows in one direction: *from* the system that gives rise to the influence *to* the object upon which the influence acts. This asymmetry is represented in causal influence graphs not just by the fact that arrowheads are placed on the lines

---

[2] More precisely, such analyses typically take counterfactuals to describe patterns of association between events across similar possible worlds, where some distinguished measure of similarity between worlds is assumed. In this case, the question of why certain counterfactuals are true becomes the question of why similar worlds share the patterns of association that they do. Or perhaps equivalently, what is the significance of the distinguished similarity metric that we use when assessing counterfactuals?

connecting variables (as is the case in SCM causal graphs), but also by the fact that arrows representing causal influence can have multiple tails but not multiple heads.

Through the CIM framework, the ontology of power and influence makes it clear why we take certain equations as fundamental, and begins to answer the question of why we tend to exclude backtracking counterfactuals. So the CIM framework has a theoretical advantage over the SCM framework. I will argue below that the CIM framework also has practical advantages.

9.6 Influence Invariance

The most important benefit of causal influence models is that they provide information regarding a type of invariance that is ignored in the SCM framework. Woodward correctly identifies invariance as a crucial feature of causal explanation, stating that 'a necessary and sufficient condition for a generalisation to describe a causal relationship is that it be invariant under some appropriate set of interventions' (2003, 15). In particular, he goes on to tell us, it is level invariance and modularity that are characteristic of causal relationships. A structural equation, he says, describes a causal relationship between variables in a model, and the equation must be invariant under interventions which change the values of the variables on the right-hand side (within an appropriate range) and invariant under interventions which change other structural equations in the model by setting their dependent variable to a given value.

In earlier chapters, I have put a similar stress on the importance of invariance, arguing that a particular kind of invariance—let us call it *influence invariance*—is essential for reductive explanation. The method of reductive explanation assumes that the influences produced within a subsystem in a given state remain invariant regardless of the larger system of which the subsystem is a part. It is this invariance that allows us to understand and predict the behaviour of complex systems based on our knowledge of the parts, and, in general, allows us to apply, in one situation, knowledge that was gained in some other situation. Without assuming invariance of this kind, each time we wanted to build a causal model of a new situation, we would have to start from scratch, unable to make use of our knowledge of similar systems.

Woodward himself recognizes the importance of this kind of invariance. Commenting on a strategy for mechanism discovery outlined by Linley Darden (2002), Woodward says:

> My notion of modularity is closely connected to what Darden calls in her paper 'modular subassembly'. Recall that this is a strategy for mechanism discovery that proceeds by 'hypothesiz[ing] that a mechanism consists of known modules or types of modules. One cobbles together different modules to construct a

hypothesized mechanism' (Darden, 2002). If this strategy is to work, it must be possible to add a new module or component to a structure consisting of other modules or to replace one module with another without disrupting or changing the other modules in the structure. This is just another way of expressing the idea that modules or components should be independently changeable.

Woodward (2002, S375)

If we identify the 'modules' mentioned here with causal mechanisms (taken to be the manifestation of an influence by a causal power), then influence invariance ensures that it is possible to add a new module or component to a structure or replace one module with another without disrupting or changing the other modules in the structure.

However, Woodward is wrong about the close connection between his concept of modularity and the strategy of modular subassembly, for neither level invariance, modularity, nor their combination, provides the kind of invariance that is required. To see this, consider again the causal model of three charged particles depicted in Figure 9.3, and let us ask what would happen if we were to intervene on the system by *adding or removing* a charged particle. This intervention is not the kind that changes the values of some dependent variables in equation 9.12, nor is it the kind that replaces some other structural equation in the model with a specification of the value of the dependent variable. In short, this is not the kind of intervention to which level invariance and modularity apply, so level invariance and modularity tell us precisely nothing about what, if anything, remains invariant under this kind of intervention. Influence invariance, on the other hand, tells us that the influences within each two-object system will remain invariant regardless of changes outside that system (and, more generally, the functional form of the influences will remain invariant).

Imagine, then, that you have established that the causal model represented in equation 9.12 and Figure 9.3 correctly describes a given system of three charged particles arranged in a line, and suppose that you now wish to construct a causal model for a system of four charged particles arranged in a line. If we are limited to the information contained in the SCM model, then our original three-particle model will be of no help at all when constructing the four-particle model. We cannot expect anything about the structural equations to be the same in the new model, nor can we expect the facts about what counterfactually depends on what to remain the same. None of the invariances assumed in the SCM model apply.

If we stick to the SCM framework, but add the assumption of influence invariance, we can do a little better. Influence invariance tells us that the influences between the original three particles will remain unchanged by the addition of a fourth particle. Since counterfactual dependencies are the result of causal influences, we can conclude that facts about what counterfactually depends on what will remain unchanged with the addition of the fourth particle. Thus, influence invariance tells us that the SCM graph for the four-particle system will contain the SCM graph of the original three-particle system. We are still left completely in the dark, however, about how the variables describing the new particle fit into this graph, and about the structural equations for the new system.

The causal influence model, on the other hand, helps answer these questions. Influence invariance tells us that the functional form of an influence depends on the state of the subsystem that produces it, and nothing else. So, if we take a complex system and add a subsystem of a type whose influences are already known, then this system will add an influence with a known functional form. Now, Figure 9.4 shows that there are two influences acting on A_3, each associated with a different two-particle subsystem. When we add a fourth charged particle to our system, we are adding three 2-particle subsystems, each of which is of the same type as the subsystems that give rise to influences in the original three-particle influence graph. Since we are not interested in the acceleration of any particle other than particle 3, two of these new subsystems exert influences on variables that are not in our model and so can be ignored. The third two-particle subsystem will exert an influence on A_3 which has exactly the same functional form as the influences in the three-particle influence graph. If we further assume that no influences are added by interactions between three or more particles (that is, if we assume that the causal powers involved do not have a rank greater than two), it follows that the four-particle graph will look like Figure 9.6.

In general, influence invariance allows us to piece together complex causal influence graphs from more simple ones. If a (possibly multi-tailed) arrow connects a group of variables in a causal influence graph, then a similar arrow will connect any group of variables of that kind, no matter how complex the system in which they appear. Thus, causal influence graphs are truly modular: each multi-tailed arrow defines a module type and new graphs can be built by arbitrarily combining such modules. The same cannot be said of the arrows in an SCM graph. The fact that there is an arrow from D_1 to A_3 in Figure 9.3, for example, does not imply that the acceleration of a particle will always counterfactually depend on its distance

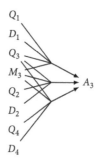

Figure 9.6. Four charged particles

from any other object (remember that the model depicted in that graph ignores gravitational effects). Influences are produced by systems of interacting objects, not by pairs of individual variables. Causal influence graphs keep track of the former; SCM graphs do not.

There is one complication when it comes to piecing together causal influence graphs from smaller modules. Our new graph may contain systems with more objects than any of the modules from which it is built, and any of these larger systems could, theoretically, involve new higher-rank powers which will manifest influences that need to be included in the graph. If, for example, we thought that the acceleration of a charged particle could be the target of a fourth-rank influence generated by the interaction between four charged particles, then we would need to include this influence in our four-particle graph. In order to construct a causal graph describing a system with more objects than any of the modules from which it is built, we must assume that the system does not generate any influences of a rank higher than that of the influences described by the modules. This is the point of Assumption 2 identified in Chapter 5, which states that there are no powers of rank greater than some number N. If this assumption is correct, then we can use modules to construct causal influence graphs for systems with more than N objects so long as we have a module for every relevant influence of rank less than or equal to N.

9.7 Implications for Structural Equations

We have seen that causal influence graphs do a much better job of capturing influence invariance than do SCM causal graphs. This fact means that we can construct complex causal influence graphs (and hence also complex SCM graphs) from simpler, well-understood causal influence graphs. But, in order to apply the full power of the causal modelling framework to a system, we need the structural equations for that system; the causal graph alone is not enough. Since the SCM framework does not even give us a way to build complex causal graphs from simpler ones, it cannot provide a way to infer the structural equations for a complex system from our knowledge of the simpler systems. Can the causal influence framework do any better?

The answer is yes. Causal influence graphs, like the SCM graphs that they imply, can tell us which variables will appear on the right-hand side of each structural equation. Once we have constructed the causal influence graph for a complex system, therefore, we will at the very least know something about the functional form of the structural equations for that system. So, for example, Figure 9.6 Tells us that the structural equation for A_3 will have the following form:

$$A_3 = f(Q_1, D_1, Q_2, D_2, Q_4, D_4, Q_3, M_3) \qquad (9.16)$$

Even this very limited information is more than we get from a knowledge of the simpler three-particle system within the SCM framework. But causal influence graphs have further implications for structural equations. According to Assumption 3 of Chapter 5, the change in a variable will be determined by the joint action of the basic influences acting on that variable. This means that the variable will only 'see' the influences directed at it, not the values of all the variables which may be responsible for those influences. Thus, a causal influence graph places an extra constraint on the form that structural equations can take. Consider, for example, the causal influence graph in Figure 9.7, which is constructed from two simpler graphs or 'modules'—one which describes the WXE system and one which describes the YZE system.

Figure 9.7. Graph with two basic mechanisms

This graph shows that there are two influences acting on the variable E, so the structural equation for E must express E as some function, call it f, of these two influences. Now, the graph also tells us that one of these influences is itself a function, call it g, of W and X, while the second influence is a function, call it h, of Y and Z. Thus, the structural equation for E must take the following form:

$$E = f(g(W, X), h(Y, Z)) \qquad (9.17)$$

To see that this places a real constraint on the equation, note that the following function of W, X, Y, and Z, for example, cannot be expressed in this form:

$$E = \frac{X}{Z} + \frac{Y}{W} \qquad (9.18)$$

Equation 9.18, therefore, cannot be the structural equation for E in a system described by the causal influence graph in Figure 9.7.

In fact, if the arguments of Chapter 5 are correct, then the constraint on structural equations is even stronger. In equation 9.17, the functions g and h characterize the two influences acting on E, while the function f computes the result of the two influences acting together. In line with the arguments of Section 5.5.2, then, we must be able to write the structural equation in the form of 9.17 with the additional constraints that g and h represent the outcomes that would occur if each influence were to occur in isolation. In other words, g must be the structural equation for E in the WXE module while h must be the structural equation for E in the YZE module.

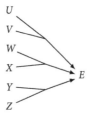

Figure 9.8. Graph with a third basic mechanism

Note further that f, g and h all output possible values for E. Thus, we can regard f as a binary operator, $\star$, on the space of possible values for E, and we can rewrite Equation 9.17 as:

$$E = g(W, X) \star h(Y, Z) \tag{9.19}$$

Finally, since the operator $\star$ represents the combined effect of the influences whose characteristic outcomes are represented by g and h, it follows that $\star$ is the same operator identified with this symbol in Chapter 5. As pointed out in that chapter, if we are to take full advantage of the reductive method, then the operator $\star$ must form a commutative semigroup over the possible values of E.

Suppose, now, that we were to add to our model a new module UVE of the same type as the WXE module. Since adding this module will not affect the existing modules, and since it is the same type as the known WXE module, we know that the graph for the new model will be that shown in Figure 9.8.

Furthermore, since the new module is of the same type as the WXE module, the influence the new module exerts will have the same functional form, and so will be directed towards $g(U, V)$. Finally, if $\star$ is a commutative semigroup, then the structural equation for E will now be:

$$E = g(U, V) \star (g(W, X) \star h(Y, Z)) \tag{9.20}$$

Thus, modules that represent causal mechanisms will be 'additive' in a general sense.

These formal constraints on structural equations have practical consequences, since they limit the search space when trying to fit a model to empirical data. If we construct a causal influence model for a complex system from simpler, well-understood causal influence models, then the task of figuring out the structural equations for the complex model becomes a task of finding a binary composition operator that fits the data.

If we already know the algebra of composition for the relevant influences then we can simply read off the structural equations for the complex system based on our knowledge of the structural equations of the simpler systems from which it is composed. The theory of Newtonian forces, for example, includes a specification of the algebra of the relevant influences (i.e. forces) and this is why it is such a useful predictive tool. We can use our knowledge of two-particle interactions to construct structural equations for arbitrarily complex systems.

9.8   An Improved Model of Intervention

A further advantage of the causal influence framework is that it can provide a richer account of interventions. In the SCM framework, interventions are thought of as modelling idealized experimental manipulations of a variable. The idealisation takes the form of assuming that (a) the manipulation is independent of any variables in the system, (b) manipulation of a variable breaks any dependence that variable might have had on other variables in the system, and (c) possible manipulations are not limited by practical or technological concerns and, indeed, may not be limited to physical possibilities.

In the real world it is not always possible, or even desirable, to perform such an idealized intervention. Suppose, for example, that you wished to intervene on the system represented in Figure 9.1 to set the value indicated by the barometer, and suppose, further, that the barometer is of the aneroid type. Aneroid barometers contain a small thin-walled chamber and work by measuring the deformation of the walls caused by a difference in pressure inside and outside the chamber. We could intervene to set the value indicated by such a barometer in a way that satisfies assumptions (a) and (b) by attaching a pair of electrically powered tweezers to the barometer in such a way that they can be used to apply force to the aneroid chamber. We could then control the reading on the barometer by varying the force exerted by these tweezers—say, by turning a dial that controls the voltage delivered to the tweezers. Assumption (a) will be satisfied so long as the setting on the tweezer dial (and the effect this has) is independent of the other variables in the model. Such independence is easy enough to arrange. To satisfy assumption (b), we need to break, or 'switch off', the influence that the atmosphere exerts on B. As I argued in earlier chapters, however, basic causal powers will always manifest their influence in the appropriate triggering conditions. In order to switch off an influence, therefore, we must interfere with its triggering conditions. Now, the triggering conditions for feeling the influence of atmospheric pressure are simply being immersed in the atmosphere. In order to switch off the influence of atmospheric pressure on the barometer, therefore, we need to remove it from the atmosphere. We can achieve this, for example, by placing the barometer (with tweezers attached) into a rigid airtight evacuated enclosure.[3] Having assured that

---

[3] We can give a more detailed explanation of what is going on here if we stop treating atmospheric pressure as a basic causal power and note that the influence of atmospheric pressure on an object is the resultant of chaining together a huge number of more basic influences. Molecules of air high in

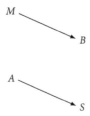

Figure 9.9. An ideal intervention on the barometer

(a) and (b) are satisfied, we could then manipulate the value indicated on the barometer to our heart's desire simply by turning the dial.

A causal influence model of such an intervention on the atmosphere-storm-barometer system is shown in Figure 9.9 (where M represents the intervention). In this case the nature of the intervention is not influenced by any systems internal to the model, and the influence of the manipulation upon a particular variable replaces (or, at least, completely overwhelms) any other influences on that variable. Thus, the intervention satisfies both (a) and (b) and matches the ideal that is modelled in the SCM framework. Indeed, because all the arrows in this particular model are single-tailed, this graph is exactly the same as the graph representing an intervention on B in the SCM model.

Now, one important reason for intervening on a variable in the real world is to measure the effect that a change to that variable has on other variables. The intervention described above, however, would not be of much use for this purpose since the intervention effectively removes the barometer from the original system altogether. By isolating the barometer to block any incoming influences we are likely to block any outgoing influences as well. For example, such an intervention would not be an effective way to establish that the readings on normal barometers have no causal influence on whether there is a storm nearby, since sticking the barometer into a rigid airtight container might plausibly shield this influence.

A more useful (and more practical) way of manipulating the value shown on the barometer would be to attach the tweezers, but otherwise leave the barometer as it is—in particular, to not place the barometer into a rigid airtight container. Once again we will be able to manipulate the value indicated on the barometer by turning the dial on the tweezers, but since our intervention on the barometer is much more minimal than in the previous case, there is less chance that we will have

---

the atmosphere exert forces on molecules below them, which in turn exert forces on molecules further down, and so on. By placing the barometer in the rigid container we interfere with this chain. Molecules outside the chamber exert forces on the container walls, but these forces are counteracted by forces between the molecules within the walls of the container, so that the resultant force on the walls is zero. Of course, if the air pressure outside the container is high enough, the forces between molecules in the walls will not be able to balance the atmospheric pressure, the container will collapse, and the baraometer will once again be (violently) subject to the influence of atmospheric pressure.

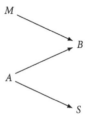

Figure 9.10. A more realistic intervention on the barometer

inadvertently blocked any outgoing influences that B might have on whether or not there is a storm. However, this intervention will not satisfy (b), because we have not completely removed the influence of atmospheric pressure on the value of the barometer. To see this, note that how hard the tweezers need to squeeze to get the barometer to indicate the desired value will depend on the atmospheric pressure. In particular, the value indicated on the barometer will still counterfactually depend on atmospheric pressure; if you fix the voltage delivered to the tweezers and vary the atmospheric pressure, then the reading on the barometer will change.

A graph of this situation is shown in Figure 9.10. In this situation, the influence exerted by our intervention does not *replace* the influence of atmospheric pressure. Rather the two influences *combine* to produce an outcome on the barometer. Many real-life interventions will work this way. It will often be impossible or impractical to simply 'switch off' the influences exerted on a given variable in the model, and when it is possible, doing so may involve such a large change to the system being studied that the intervened-upon system cannot tell us anything useful about the causal relations within the original system (as was the case with the first intervention on the barometer described above). Thus, if we want to predict what will happen when we intervene in a real system, it will be important to be able to model interventions that have this structure in which the intervention combines with existing influences.

Note that even though this more realistic intervention does not switch off the influence of A on B, and so does not satisfy (b), the intervention can still be used to distinguish correlations based on common causes from those based on direct causal links. So long as A stays constant, if we wiggle M and find that B wiggles but S does not, then we can conclude that the correlation between B and S is due to a common cause, as shown in Figure 9.10. But if S also wiggles when we wiggle M, then we can conclude that there is a causal link from B to S. The situation is muddied if A is not constant, but so long as the way A changes is significantly different from the way we change M (for example, if A changes much more slowly than M), we should be able to distinguish the results of our intervention, and so distinguish common causes from direct causes.

Even when we cannot switch off the influences acting upon a variable, it is sometimes possible to remove the counterfactual dependence of that variable on other variables in the model. Consider again the second intervention on the barometer described above. Suppose that we wish to be able to set the value indicated by the barometer by turning the dial to a value that is not dependent on atmospheric pressure. So, for example, if we wish the barometer to read '900 hPa', we want to be able to simply turn the dial to a point labelled '900', and we want this to work regardless of the atmospheric pressure. We could achieve this by setting the system up to automatically adjust the amount of force applied by the tweezers in response to the atmospheric pressure. So, for example, when you turn the dial to 900, the system measures the atmospheric pressure and then calculates and applies the force that, when combined with atmospheric pressure, results in the barometer reading 900. In this case, we have satisfied (b) by removing the counterfactual dependence of B upon A (but not the causal influence of A on B), but we have done this by violating (a). Since the force exerted by the tweezers depends on the atmospheric pressure, the nature of the intervention upon the barometer now depends upon one of the variables in the model (namely A). In general, when we cannot switch off the influences acting on a variable, we will only be able to intervene on that variable in a way that satisfies (a) if the intervention violates (b) and vice versa.

A graph of this third situation is shown in Figure 9.11. It is important to note that the arrows in this graph represent causal influences, not relations of counterfactual dependence. As described above, it is possible to fine-tune the intervention in such a way that it removes the counterfactual dependence of B on A without removing the causal influence of A on B. The SCM framework is incapable of distinguishing this situation from one in which the intervention does remove the causal influence of A on B.

A perusal of the three intervention graphs discussed above suggests that there is a fourth possibility, which is shown in Figure 9.12. We could perform such an intervention by attaching tweezers to the barometer and placing the set-up into a rigid airtight container to remove the influence of the atmosphere, as in the first case above. We could then adjust the dial on the tweezers in response to the

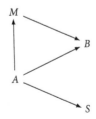

Figure 9.11. A compensated intervention on the barometer

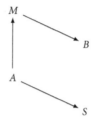

Figure 9.12. A fourth intervention scenario

atmospheric pressure, say, by turning it to match the reading of a second barometer which is not isolated from the atmosphere. I am not sure why we might want to do this but it would result in an intervention matching the graph in Figure 9.12. Like the third case above, this intervention fails to satisfy (a), since the nature of the intervention itself is influenced by A. Also like the third case above, the intervention satisfies (b), since it removes any counterfactual dependence of B on A (though in this case it achieves this by removing any causal influence of A on B). So, because the SCM framework keeps track only of counterfactual dependence and not causal influence, it cannot distinguish this fourth possible intervention from the third case above.

To model interventions in the real world, we need to be able to model not only the ideal situation shown in Figure 9.9, but also the situations shown in Figures 9.10, 9.11, and possibly also 9.12. These last three possibilities can be handled easily within the causal influence framework but they cannot be easily handled within the SCM framework. The SCM framework simply *defines* interventions such that they match only the first case. But even if we relaxed this definition, the SCM framework would run into trouble, since modifying the original model of the un-intervened-upon system to model these interventions involves adding and combining new influences. That is to say, we must model the original system as part of a larger system. As we have seen above, however, the SCM model gives us no tools for doing this (other than in the idealized case of an intervention of the first type). Thus, I suggest, the causal influence framework is a better tool for modelling intervention than the SCM framework.

9.9 Conclusion

In this chapter, I have presented the causal influence modelling framework, and argued that it has several advantages over the traditional structural causal modelling framework. First of all, the CIM framework subsumes the SCM framework, and so everything that can be done within the SCM framework can also be done within the CIM framework. The CIM framework has two big practical advantages over the SCM framework. First, because the CIM framework explicitly identifies basic mechanisms which satisfy influence invariance, it is truly modular. These basic mechanisms can be mixed and matched at will to produce new causal models from old. All that is required to do this is knowledge of the algebra of composition for the influences involved. Even if the algebra of composition is not known, the modular nature of the basic mechanisms represented in the CIM framework puts constraints on the form of the structural equations involved. Recognizing these constraints will simplify the task of constructing a causal model from scratch. The second practical advantage of the CIM model is that it can model non-ideal interventions. Since most real-world interventions will not be ideal, the ability to model these non-ideal interventions is extremely useful.

The CIM framework also has a theoretical advantage. Like the standard SCM framework, the CIM framework provides a systematic way to pose and answer questions about relations of counterfactual and probabilistic determination. Unlike the standard framework, the CIM framework represents basic mechanisms and identifies these mechanisms with the manifestation of influence by causal powers. As such, the CIM framework shows precisely how relations of probabilistic and counterfactual determination can be grounded in an ontology of power and influence. The CIM framework, therefore, fulfils the promise made in the previous chapter that claims about causation as dependence can be understood as being claims about causal powers and the influences that they manifest. Ultimately, then, talk of causation as production and talk of causation as dependence can be understood as having the same subject matter: causal powers and causal influence.

10

Emergence and the Failure
of Reduction

The great success of the reductive approach to explanation suggests that, in many cases at least, the assumptions of the approach are satisfied. But there are cases in which the reductive explanation has not (yet) been successful. Phenomena associated with consciousness, biology, economics, crowd behaviour, and even fluid dynamics, for example, have proved resistant to reductive explanation. What, if anything, do these cases tell us about the world? This is the question that will occupy us in this chapter.

The failure of reductive explanation is closely linked to the much-discussed concept of *emergence*. The last few decades have seen a resurgence of interest in the concept of emergence—both within philosophy and within the sciences.[1] What these discussions have in common is the vague idea that emergence has to do with complex systems in which the whole is 'greater than the sum of the parts'. Beyond this core notion, however, there is a great deal of variation in what people mean by emergence. Indeed Jaegwon Kim commented that those discussing emergence, more often than not, talk past each other. Sometimes, he says, 'one gets the impression that the only thing that the participants share is the word 'emergence'.' (Kim, 2006b, 548) In this chapter I will argue that there is a close connection between the concept of emergence and the failure of reductive explanation. In particular, I will argue, the term 'emergence' is appropriately applied in situations in which the assumption of supervenience holds, but reductive explanation fails. Understanding emergence this way, I believe, will give us the tools we need to better understand the concept, and allow us to catalogue different ways in which emergence may come about. Understanding emergence this way will also give us the tools we need to defend the possibility of truly ontological emergence from a suite of influential arguments that Kim has directed against this possibility.

Before we consider the connection between emergence and reduction, though, we can do a little more to unpack the idea of a whole that is greater than the sum of its parts. According to Timothy O'Connor and Hong Yu Wong (2015), there are two core claims which characterize emergence. First, emergent entities 'arise' out of more fundamental entities. Second, emergent entities are, nonetheless, 'novel' or 'irreducible' with respect to those fundamental entities. The first of these claims

---

[1] For some overviews and edited collections, see Beckermann et al. (1992); Holland (2000); Clayton and Davies (2006); Kistler (2006); Corradini and O'Connor (2010).

captures the idea that we are dealing with a whole which is composed of parts; the second captures the idea that this whole is somehow greater than the mere sum of its parts. In the next section I will consider the way in which emergent entities are supposed to 'arise' out of more fundamental entities, and argue that this is closely connected to the assumption of supervenience which underlies reductive explanation. The remainder of the chapter will employ our understanding of reductive explanation to shed light on the ways in which an emergent entity may be said to be 'novel' or 'irreducible'.

10.1 Emergence and Supervenience

Emergentists have always insisted that emergentism is not a kind of substance dualism. The classical British emergentists (Mill, 1843; Bain, 1870; Lewes, 1874; Alexander, 1920; Morgan, 1923; Broad, 1925) were particularly concerned to distance their view from that of substantial vitalism (Broad, 1925, 58), while modern emergentists are more concerned with a dualism of the Cartesian variety (O'Connor and Wong, 2005). But in either case, the idea is that the world is composed from the stuff of microphysics—nothing else is 'added from outside', as Tim Crane (2001) puts it. Emergentism, then, is committed to some kind of physical monism. Following James Van Cleve (1990), this commitment is often cashed out in terms of supervenience (recall that properties in the set A supervene on properties in the set B iff there can be no difference in an object's A-properties without a difference in the object's B-properties). In particular, emergentism is usually said to involve the following claim:

Emergentist Supervenience The emergent properties of a whole supervene (with at least nomological necessity) upon the properties of its microphysical parts.

Wong (2010) has pointed out that supervenience of this kind merely establishes a necessary correlation between emergents and their microphysical emergence bases; it does not rule out the possibility that this is a correlation between physical properties and properties of non-physical substances or the like. Descartes, for example, believed that there were causal laws connecting the non-physical mind and the physical brain. In his *Meditations* he tells us that 'each of the movements which are in the portion of the brain by which the mind is immediately affected brings about one particular sensation only' (Descartes, 1911, 197). Now, it seems consistent with Descartes's position that a sensation of this kind can *only* be produced by an appropriate 'movement' in the brain, in which case, if you have a certain sensation, you must (with nomological necessity) have the appropriate brain state. If this were the case, then Cartesian mental states would supervene on physical brain states. Thus, supervenience is, in fact, compatible with dualism.

Wong concludes that the desire to avoid dualism cannot legitimately motivate the claim of supervenience (Wong, 2010, 16). Wong's reasoning here is flawed, however. It is not because supervenience is thought to imply the denial of dualism that it is used to characterize emergence. Rather, supervenience is said to be characteristic of emergence because it is *implied by* the kind of physical monism that is characteristic of the emergentist's denial of dualism. Crane, for example, argues as follows:

> For if they do not thus supervene, then it seems somewhat perverse to describe the properties as 'emergent'. Presumably part of the point of this label is to pick out the sense in which putting a thing's parts together gives you something new—but not because you have 'added' something 'from the outside'. If emergentism is to be distinguished from dualism and vitalism (which do add something 'from the outside') then it must reject this strong notion of emergence. The upshot is that a reasonable emergentist thesis is committed to the supervenience of a whole's properties on the properties of its parts.   (Crane, 2001, 212–13)

Emergentist supervenience is not erroneously thought to be justified because it implies physical monism. Rather, it is justified because it is implied by the the emergentist's prior commitment to physical monism.

Now, recall Assumption 1 of reductive explanation:

Supervenience   When attempting a reductive explanation of some phenomenon P in complex system S, we assume that there is a set of basic objects of S such that the relevant properties of P supervene on the properties of, and relations between, these basic objects.

If, as the emergentist claims, the emergent properties of a whole supervene on the properties of, and relations between, its microphysical parts, then emergent properties will satisfy the assumption of supervenience that underlies reductive explanation. Since the emergentist's supervenience claim is, itself, implied by physical monism, it follows that a world in which reductive explanation's assumption of supervenience fails is a world in which physical monism is false; it is a world with properties or substances that are completely independent (both logically and nomologically) of the physical state of a system.

Note that the converse is not true: a world in which physical monism is false is not necessarily a world in which the supervenience assumption of reductive explanation fails. As Wong pointed out, emergentist supervenience (and hence the supervenience required by reductive explanation) could hold in a dualist world if there is a lawful correlation between mental states and physical states. More interestingly, the supervenience assumption of reductive explanation (though not emergentist supervenience) could be satisfied in a world in which there are complex mental systems whose properties supervene on the properties of, and relations between, mental components. Thus, it is possible that reductive explanation could be employed to explain non-physical systems, if there are any. For example, folk psychology attempts to explain people's behaviour in terms of their beliefs, desires, emotional states, and the like. What is more, beliefs, desires, and emotional states can, arguably, be understood as dispositions that operate together to produce a person's behaviour (both behaviourism and functionalism seem to imply such a view). Thus, folk psychology can be understood as a rudimentary attempt at a reductive explanation of human behaviour, an attempt that takes mental properties such as a person's beliefs, desires, and emotional states as basic. Importantly, the success or otherwise of this attempt does not seem to depend on whether or not the mental properties that are taken as basic supervene on physical states.

The possibility of non-physical reductive explanation suggests a generalization of the concept of emergence. Suppose that we have so far been successful at reductively explaining the behaviour of complex mental systems in terms of the powers of a set of basic mental component entities. Suppose, further, that we encounter a complex mental system that resists such reductive explanation in terms of its 'micro-psychological' components. It would seem that there are two possible ontological explanations for this failure.[2] It may be that the properties of the complex mental system do not supervene on the properties and relations between its components. That is to say, it may be that something has been 'added from outside' the psychological system of basic mental states. Alternatively (if emergence is indeed a possibility), it may be that there are some properties of the complex mental system that are 'novel' or 'irreducible' despite the fact that the properties of the complex system supervene on those of its components. So, if emergence from microphysics is a possibility, it seems that something like emergence from micro-psychology should also be a possibility.

I suggest, therefore, that we generalize the concept of emergence by dropping the claim of physical monism that characterized classical British emergentism, and replace it with the following generalized supervenience thesis:

Emergentist Supervenience[*] The emergent properties of a whole supervene (with at least nomological necessity) upon the properties of its parts.

This supervenience claim allows for the possibility of emergence in non-physical systems. It would even allow for emergence in hybrid systems that have both physical and non-physical parts (if such systems are possible). Thus, I believe, this generalized supervenience claim is the appropriate way to understand the first core claim of emergentism.

---

[2] A third, non-ontological possibility is that the failure is due to our own epistemic limitations; see below.

10.2 Epistemic Emergence and Ontological Emergence

What of the second core claim of emergentism, that the whole has properties that are 'novel' or 'irreducible'? Different ways of cashing out the concept of novelty involved here have led to a bewildering variety of accounts of emergence, and it is this variety that prompted Kim's claim that participants in the debate are often talking past each other. However, these accounts can be classified into two types: epistemic accounts of emergence, which characterize the novelty of emergent properties epistemically, and ontological accounts of emergence, which characterize this novelty ontologically (this distinction is similar to, but not quite equivalent to, the distinction that David Chalmers (2006) draws between *weak* and *strong* emergence).

Consider O'Connor and Wong's definition of epistemic emergence, which they state in terms of prediction:

> Emergent properties are systemic features of complex systems which could not be predicted (practically speaking; or for any finite knower; or for even an ideal knower) from the standpoint of a pre-emergent stage, despite a thorough knowledge of the features of, and laws governing, their parts.
>
> (O'Connor and Wong, 2015)

The parenthetical specifications list three different possible strengths, or grades, of epistemic emergence. Now, as Kim (1999, 5) points out, the classical British emergentists spoke both of the *unpredictability* and of the *inexplicability* of emergent properties. If we generalize O'Connor and Wong's definition to include the failure of explanation, then, we get the following, slightly more general account of epistemic emergence:

Epistemic Emergence Emergent properties are systemic features of complex systems which cannot be predicted or explained (practically speaking; or for any finite knower; or for even an ideal knower) from the standpoint of a pre-emergent stage, despite a thorough knowledge of the features of, and laws governing, their parts.

Epistemic emergence, thus characterized, is nothing other than a failure (practically speaking; or for any finite knower; or for even an ideal knower) of reductive explanation (together with the satisfaction of Emergentist Supervenience*). For what has failed in such a situation is an attempt to explain or predict the complex system using our knowledge of the properties of, and laws governing, its parts *from the standpoint of a pre-emergent stage*. The reference to the standpoint of a pre-emergent stage here stipulates that the knowledge we are making use of is a knowledge of the properties and laws governing the parts as they are *before they are assembled into the whole* in question. This attempt to understand a complex system using what we know of the parts when separated from the system is precisely what defines the reductive approach to explanation.

Epistemic emergence need not be anything mysterious; there is nothing strange about the idea that a system might be too complex for us to understand in terms of its parts. Ontological accounts of emergence, on the other hand, do have an air of mystery about them. For an ontological account of emergence attempts to describe—from a God's eye view—a way in which a complex system could have a genuinely 'novel' property which nonetheless supervenes on the properties of, and relations between, its parts. Indeed, Kim (1989; 1990; 1992; 1993a; 1993b; 1998; 2003; 2005; 2006a; 2006c; 2006b) has argued that the concept of a property that supervenes on, but is not reducible to, the properties and relations of a system's parts is not merely mysterious, but incoherent. I believe, however, that our understanding of reductive explanation will give us the tools to demystify ontological emergence and provide an account of such emergence that is recognizably continuous with the emergentist tradition, yet is immune to Kim's arguments.

The first step in developing an account of ontological emergence is to recognize that ontological emergence implies epistemic emergence. The difference between the two kinds of emergence is that ontological emergence is not *merely* epistemic. Ontological emergence is simply something in the world that could give rise to epistemic emergence independently of our epistemic limitations. Now, I argued above that epistemic emergence occurs in cases where reductive explanation fails even though the assumption of supervenience holds. So ontological emergence will describe cases where something in the world leads to the failure of reductive explanation independently of our epistemic limitations. To better understand ontological emergence, and its relation to reductive explanation, therefore, it will be useful to consider how it is that the reductive approach could fail.

Suppose that we wish to provide a reductive explanation of some phenomenon in terms of a set of basic entities, B. The discussion of reductive explanation in previous chapters tells us that such an explanation would involve attributing a set of basic powers to these basic entities, formulating the laws for composing the influences exerted by these basic powers, and showing that these powers and composition laws would give rise to the phenomenon to be explained. Suppose, then that we have attempted a reductive explanation of some phenomenon in terms of a set of basic entities, B, and that our attempts at such explanation have failed. This failure may be due to epistemic problems—we may have left an important entity out of B, we may have misunderstood the properties of the enties in B, or the situation might just be too complex for us to succesfully predict the outcome of all the interactions within B. But another possible explanation for our failure is that the system under consideration does not satisfy the assumptions of reductive explanation. Now, whether or not the assumptions of reductive explanation are satisfied is a feature of the world that is independent of our epistemic limitations, so it seems fair to say that a failure of reductive explanation due to the falsity of one of its assumptions is an *ontological failure* of the reductive approach. My suggestion, then, is that cases of ontological emergence simply are cases of ontological failure of reductive explanation. In particular, we get ontological emergence when the assumption of supervenience is satisfied, but one or more of the other assumptions of reductive explanation are not.

If all of the assumptions of reductive explanation are satisfied, then a reductive explanation should, in principle, be possible. Thus, any failure of the method that is not due to the failure of one or more of those assumptions, must be attributable to our own poor choices or epistemic limitations. Even if a reductive explanation is possible, and we have chosen the correct set of basic components to produce such an explanation, our attempts can fail. This would be the case if we could not isolate the basic entities sufficiently to study their causal powers, or if we cannot figure out the laws for composing influences, or if there are just so many components involved that the calculations are unmanageable. When reductive explanation fails, but this failure is not ontological, let us say that the failure is *merely epistemic*, or that we have a case of *merely epistemic emergence*.

In practice, it will often not be possible to tell whether a given failure of reductive explanation is ontological or merely epistemic. In situations where B contains more than just a few entities, the possible explanations rapidly become complex enough that there is always the possibility that our attempts at reductive explanation have failed due to our own inability to handle such complexity. Nonetheless, if genuine attempts at reductive explanation repeatedly fail, then it seems worth considering the possibility that the failure is ontological. This is especially the case if the causal powers of the basic entities are well understood—as evidenced, say, by their use in successful reductive explanations of other phenomena. For if the basic entities and powers involved are well understood in other contexts, this lowers the likelihood that our failure is due to a lack of understanding of the powers involved, or of the laws for composing influences. The most likely explanations for failure are then either that the system is too complex or that the failure is ontological.

My suggestion, then, is that ontological emergence occurs when the supervenience assumption of reductive explanation is satisfied but one or more of the other assumptions is not. If this is indeed the right way to understand ontological emergence, then it follows that there may be different types of ontological emergence corresponding to the failure of different assumptions. If we suppose that the supervenience assumption is satisfied, then there are three other independent assumptions of reductive explanation and the reductive method (recall that Assumption 3 (Changes Determined by Influence) is implied by, and so not independent of, Assumption 4). Thus, we might think that there are three distinct types of ontological emergence. In the remainder of this chapter, I will consider these three assumptions and argue that the failure of the third is not really the kind of thing that would legitimately be regarded as emergence. The failure of the first two, however, does lead to phenomena that one could characterize as emergent. The failure of Assumption 2 (that the relevant powers have a limited rank), in particular, seems to be associated with traditional concepts of emergent properties, and I will spend the majority of the chapter developing this thought and defending the possibility of such emergence from Kim's argument against ontological emergence. The failure of Assumption 4 (regarding the algebra of influence composition), on the other hand, seems to give rise to a novel form of emergence associated with the composition of influences.

10.3 Assumption 2 and High-Rank Powers

In Chapter 5, I defined a power to be of *rank n* (relative to a given set of basic entities) iff n is the cardinality of the smallest system with respect to which the influence that the power exerts is invariant. That is to say, a power is of rank n just in case n basic entities are required to determine whether, and in what manner, the power manifests. So, for example, electric charge is a second-rank power, since the influences it produces act between pairs of charged particles in such a way that the magnitude and direction of the electromagnetic influence one charged particle exerts on another is determined by the charges of the two particles, the distance between them, and nothing else. Thus, information about a two-particle system is both necessary and sufficient in order to determine the electromagnetic influences one charged particle has on another.

In that chapter we also saw that any given reductive explanation has a rank, which I defined as equal to the rank of the highest-ranked basic power that is referenced by the explanation. When attempting a reductive explanation, therefore, we will be attempting an explanation of a particular rank, N, and in doing so we must assume that we can ignore the possibility of basic powers with a rank higher than our chosen value N. This is Assumption 2 of reductive explanation. If we have only studied the influences at play in two-particle systems, for instance, and wish to use this knowledge to explain the behaviour of a multi-particle system, then we must assume that there are no (significant) basic influences added by powers with rank higher than two. If this assumption regarding the rank of the significant basic powers in a system is not satisfied, then the attempted reductive explanation will fail.

Let us refer to any basic causal power that has a rank greater than the relevant value of N as a *high-rank power*. Explanations in physics and chemistry tend to assume that $N = 2$, so in the context of explanations which take entities from physics or chemistry as basic entities, powers with rank greater than two are likely to be classed as high-rank powers.

High-rank powers, I believe, are exactly the kind of thing that is implicated in traditional notions of ontological emergence. Consider, for example, Hong Yu Wong's definition of ontological emergence:

> Ontological emergence is the thesis that when aggregates of micro-physical properties attain a requisite level of complexity, they generate and (perhaps) sustain

> emergent natural properties. What is constitutive of ontological emergence is the novel causal influence of emergent natural properties. One significant question is how to understand the notion of novel causal influence. An intuitive gloss is that an emergent property provides a causal contribution that goes beyond causal contributions made by any of the lower-level properties had by the system and its parts taken either in isolation or in combination.   (Wong, 2010, 8)

Wong's description here is satisfied perfectly by high-rank powers. Whether, and how, a high-rank power manifests an influence is determined by the state of a system with more than N basic entities. Thus, a high-rank power can only manifest in a system with more than N basic entities, and the influence, or 'causal contribution' manifested by a high-rank power will be additional to, and so 'go beyond' the contributions made by any of the low-rank powers.

Similarly, high-rank powers would qualify as 'strongly emergent' in the sense defined by Jessica Wilson:

> Token apparently higher-level feature S is strongly emergent from token lower-level feature P on a given occasion just in case, on that occasion, (i) S broadly synchronically depends on P, and (ii) S has at least one token power not identical with any token power of P.   (Baysan and Wilson, 2017, 59)

High-rank powers can also help us make sense of the classical British emergentist distinction between 'emergent' and 'resultant' properties. Consider, for example the following statement by C. Lloyd Morgan:

> The concept of emergence was dealt with (to go no further back) by J. S. Mill...The word 'emergent', as contrasted with 'resultant', was suggested by G. H. Lewes...Both adduce examples from chemistry and from physiology; both deal with properties; both distinguish those properties (a) which are additive and subtractive only, and predictable, from those (b) which are new and unpredictable.   (Morgan, 1923, 2–3)

If we understand emergent properties to be high-rank powers, then there is a very clear and precise way of making sense of this distinction between 'emergent' and 'resultant' properties. The Earth, for example, exerts a force of 0.1 N on any 1 kg mass that is in close proximity. This power is not shared by any of the Earth's microphysical components, so there is a sense in which this power is something over and above the powers of the microphysical components. Intuitively, however, this property is not emergent, since the gravitational force of the Earth just is the vector sum of the forces contributed by all the components. In Morgan's words, it is 'additive'. This intuition can be made precise using the language developed in Chapter 6: The gravitational power of the Earth is a resultant power and this is why it does not qualify as emergent.

In general, as discussed in Chapters 5 and 6, influences can be 'added' in much the same way that forces can (though the algebra of composition need not be that of vector addition). Thus, given an explanatory basis, we can draw a distinction between basic influences, which are manifested by powers within the explanatory basis, and resultant influences, which are constituted by the joint operation of basic influences. Suppose, then, that in a given situation some complex entity, E, exerts an influence, I. If I is constituted by the composition of influences that are manifested by powers of rank lower than N, then I will be a resultant influence in our sense, and it will be predictable given an understanding of the low-rank powers and the laws of composition. Thus, the power to manifest I will count as a resultant property in Morgan's sense. Suppose, however, that I is (or is partly constituted by) an influence that is manifested by a basic high-rank power. In this case I will not be the resultant of influences manifested by low-rank powers and, as such, it will not be predictable from an understanding of the low-rank powers and the laws of composition. Thus, it makes sense to speak of the power to manifest I as an emergent property, and contrast it with merely resultant properties.

10.3.1 Case study: A Rank Three Force

To get a better idea of what a high-rank power might look like, let us consider an example of a fictitious third-rank power. Suppose that there is a charge property whose power to exert forces is as follows: any charged particle a, in the presence of two other charged particles, b and c, will feel a force directed towards the midpoint of the line between particles b and c of strength $f_{abc} = (q_a q_b q_c / r_1 r_2) \sin \theta$, where q_x is the electric charge of particle x, r_1 is the distance between b and c, r_2 is the distance between a and the midpoint of r_1, and θ is the angle between r_1 and r_2, as shown in Figure 10.1.

The function which describes how this power manifests makes essential reference to properties of all three particles, so the power this function represents is a third-rank power. The question, then, is whether the power is basic. If so, it would count as emergent, if not, it will be merely a resultant.

Now, I could simply stipulate that, in the imagined world, the power described above is basic, but such stipulation is not necessary, since I have chosen the

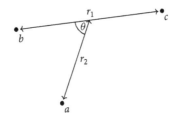

Figure 10.1. A three-particle system

form of this power such that it is logically impossible for the force manifested by the particular power to be the resultant of forces produced by two-particle interactions. To see this, suppose that the force felt by a was simply a result of combining the force exerted on a by b and the force exerted on a by c. If these forces are the result of second-rank powers, then the nature of the force exerted on a by b will be determined by the state of the two-particle subsystem consisting of a and b. Now, the relevant state of this subsystem is completely described by three independent variables, q_a, q_b, and the distance r_{ab} between a and b (I am ignoring velocities, accelerations, and the like, since our high-rank power pays no attention to such properties). Thus the force exerted on a by b must be a function of these three variables. Similarly, the force exerted on a by c will be determined by the variables q_a, q_c, and r_{ac}. Thus, the result of combining these two forces can depend on no more than five independent variables: q_a, q_b, q_c, r_{ab}, and r_{ac}. The function that describes our new force, however, takes six independent variables as arguments. Hence, the new force cannot be equivalent to a sum of forces produced by second-rank powers.

Since f_{abc} cannot be the resultant of forces produced by powers with a rank of two or lower, it will not be predictable from a knowledge of lower-rank powers and the laws for composing influences. Thus, the power to manifest f_{abc} will count as an emergent property in Morgan's sense. However, our framework allows us to say more than this, since we can distinguish two distinct ways that such an emergent property may be constituted. This distinction is worth investigating, I believe, since it may help us understand the relation between the picture I am putting forward here and other pictures of ontological emergence that have been suggested in the literature.

The argument above proves that f_{abc} cannot be the result of powers of rank two or less (and hence that there must be high-rank powers at play), but it does not follow that the force described by f_{abc} must be basic, since it may be the resultant of forces which themselves are of a rank higher than two. In fact, the symmetries of the situation, together with the assumption that space is isotropic, suggest that there are only two possibilities: f_{abc} is either a basic influence or the resultant of two high-rank forces, one directed along the line ab and the other directed along the line ac. These two possibilities are represented in Figure 10.2.

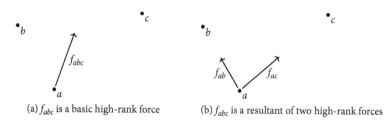

(a) f_{abc} is a basic high-rank force   (b) f_{abc} is a resultant of two high-rank forces

Figure 10.2. Two possibilities for f_{abc}

In Figure 10.2a there is a single basic force on a directed towards the point midway between b and c. A natural way of understanding this situation is that the particle pair $\{b, c\}$ has a basic causal power to attract a. The power is emergent in the traditional sense since it is possessed by a complex entity $\{b, c\}$, and is not reducible to, nor predictable from, the powers that the components b and c have in isolation. Indeed, the force f_{abc} is an example of what Brian McLaughlin calls a *configurational force*. Configurational forces, McLaughlin says, are '*fundamental* forces that can be exerted only by certain types of configurations of particles, and not by any types of pairs of particles' (McLaughlin, 1992, 52). f_{abc} fits this description in so far as it is a basic influence (and so fundamental with respect to the explanatory basis being employed) and it is a function of the configuration of a system of *three* particles. McLaughlin suggests that classical British emergentism implies the existence of configurational forces, at least if we restrict our attention to the 'framework of forces'. Similarly, within the framework of forces (where we are only concerned with powers that manifest forces), the existence of high-rank powers also implies the existence of configurational forces. This is another reason to believe that the ontological emergence generated by high-rank powers is continuous with traditional notions of ontological emergence.

The concept of high-rank powers can also help us make sense of what ontological emergence might look like if we go beyond the framework of forces. For, in general, high-rank powers will give rise to what we might call *configurational influences*: basic influences that are the manifestations of high-rank powers, and are, therefore, functions of the configuration of systems of more than two objects. I have chosen an example in which the influence manifested by the high-rank power is a force, since we have a precise mathematical way of representing forces. This makes it easy to see exactly how the high-rank power works and to confirm that there is nothing incoherent about the possibility of such a power. But there does not seem to be any reason why there could not be high-rank powers that produce other kinds of influence.

The possibility depicted in Figure 10.2a also seems to bear some resemblance to the view, put forward by Paul Humphreys (1997), that emergence involves the 'fusion' of the properties of the components to form new properties that have new causal powers. For, in this situation, the properties of a and b could be said to fuse together in the entity $\{b, c\}$ so as to produce a basic power to manifest the force f_{ab}. There is, however, a big difference between the possibility suggested here and Humphreys's idea of fused properties. For Humphreys believes that when two or more properties fuse to form a new property, the original property instances no longer exist as separate entities and they no longer have their lower-level causal powers. The fusion represented in Figure 10.2a, in contrast, is perfectly compatible with the continued operation of lower-level powers: f_{abc} can simply be added to whatever second-rank forces there may be between a and b, on the one hand, and a and c, on the other. Indeed, the invariance of causal powers implies that lower-rank powers must continue to operate even in the presence of high-rank 'fused' powers.

Figure 10.2b depicts the possibility in which the force f_{abc} is a resultant, composed from two separate forces f_{ab} and f_{ac}, the first directed along the line ab, the second directed along ac. A natural way to interpret this situation is that b has the power to attract a, as does c. But f_{ab} and f_{ac} cannot be second-rank forces, since the strengths of these two forces will need to be coordinated in order that their result is described by f_{abc}. In particular, the strength of the force that b exerts on a must depend in part on the position and charge of c in addition to a and b, and so b's power to attract a will be at least rank three. Similarly, c's power to attract a will be at least rank three.

This second possibility, in which f_{abc} is a resultant, seems to be precisely the kind of situation that Sydney Shoemaker (2002) believes must be involved in emergence. On Shoemaker's view, emergence, were it to exist, would involve microphysical entities having causal powers that remain latent until those entities are arranged in 'emergence engendering' ways. In our example, the charge q_b gives b the power to attract a, but this power remains latent until b is paired with another particle of the right kind, such as c.

O'Connor (2000, 113) has questioned the coherence of Shoemaker's picture, asking 'what could explain the responsiveness of micro-level behaviour to macro-level circumstances?' In the context of our example, O'Connor's question translates as 'Why should the force f_{ab}, that b exerts on a, depend on the properties of some third particle c?' The answer, in this situation, is simply that this is the nature of the third-rank power that b possesses. The situation may seem odd, but it is at least logically consistent.

I have argued that high-rank powers are just the kind of thing that could play the role of emergent properties as understood by the classical British emergentists. I have also suggested that high-rank powers can play the role of emergent properties in the seemingly very different pictures put forward by Shoemaker and (to an extent) Humphreys. I think it is fair to say that high-rank powers, were they to exist, would count as emergent powers.

In the next section I will show that this understanding of emergent properties as high-rank powers provides the resources to defend the possibility of emergence against influential arguments put forward by Jaegwon Kim. But before I move on, I want to make one more important point. Even though the existence of high-rank powers can lead to a failure of an attempt at reductive explanation, the existence of high-rank powers does not actually rule out the possibility of reductive explanation. When giving a reductive explanation, we must assume that there are no powers with a rank higher than some value N. If the system contains powers with a rank higher than N, then our attempt will fail. But if our attempts at reductive explanation fail, there is nothing stopping us from trying again with a higher value of N. So, for example, if we were to discover the existence of a third-rank force like that described above, we could still explain the behaviour of a complex system by adding together the influences exerted by the components. It is just that the relevant components would include three-particle subsystems as well as two-particle subsystems. Thus, high-rank emergent properties can figure in reductive explanations.

10.4 Kim's Challenge

Ontological emergentism attempts to chart a course between reductive physicalism, on the one hand, and dualism, on the other. Kim (1989; 1990; 1992; 1993a; 1993b; 1998; 2003; 2005; 2006a; 2006b; 2006c), however, has repeatedly argued that emergentism of this sort is logically impossible. Attempting such a course, he might say, is like attempting to pass between Scylla and Charybdis—there simply is no navigable space between them, and one cannot successfully avoid one without falling into the clutches of the other (and just as Circe advised Odysseus to brave Scylla rather than Charybdis, Kim urges us to opt for reductive physicalism). Now, I have just argued that traditional notions of emergence involve high-rank powers. In particular, emergence (in the traditional sense) occurs when Assumption 1 (Supervenience) of reductive explanation is satisfied, and Assumption 2 (Limited Rank) is not. So, if Kim is right that emergence is not possible, then it follows that if Assumption 1 is satisfied, Assumption 2 must be satisfied also. This would be a nice result, since it would show that there are fewer independent assumptions made by reductive explanation than I thought, and hence that we are more justified in using the method than we may have thought. Unfortunately, Kim's arguments against ontological emergence fail. Indeed, viewing emergent properties as high-rank powers in the way I have suggested above gives us the clarity to see exactly how Kim's arguments fail.

Kim's arguments against emergence involve two steps. This first step is an argument to the conclusion that all emergent causal powers must ultimately be powers to exert 'downward causation'. That is, they must be powers to exert causal influence on lower-level entities. I have no quibble with this first step of Kim's argument and will not discuss it further (but see Crisp and Warfield, 2001, and Wong, 2010, 11, for objections). I will note, however, that even if there is something wrong with the detail of this first step, downwards causation, in at least some cases, is something that will be independently accepted by almost all emergentists. Anyone who believes that mental properties are emergent, for example, is likely to accept downwards causation, for surely my intention to raise my arm (a mental—and therefore emergent—state) can be the cause of my arm raising (a physical state).

Kim's second step is to argue for the claim that downward causation is not possible. If this argument is successful, then it would follow that emergent properties, if they exist, must be causally inert—a position at odds with classical emergentism. Kim has actually provided many arguments for this second step over the years. However, they are all variations on just two distinct arguments, which I call *the argument from closure*, and *the direct argument*. I will consider these arguments in turn below.

10.4.1 The Argument from Closure

The argument from closure relies on the following principle (Kim, 2006a, 199):

Closure: If a physical event has a cause, it has a physical cause. And if a physical event has an explanation, it has a physical explanation.

Kim sees this principle as problematic for emergentism because he regards emergent properties as non-physical. To see why this is problematic, suppose that some emergent property M is a putative downward cause of some physical event P^*. *Closure* implies that P^* has a physical cause, and so if M is not physical, then there is some P, distinct from M, which causes P^*, as shown in Figure 10.3. Thus, we must either conclude that P is the cause of P^* and M is not, or conclude that P^* is overdetermined by P and M (in a way that seems objectionable). Either way, M has not brought with it any new causal powers.

The argument from closure clearly relies on the claim that emergent properties are not physical, so we could avoid the argument by simply denying this assumption and insisting that emergent properties are indeed physical. Such a position is quite compatible with the spirit of emergentism, since emergentism is traditionally intended to be a form of physical monism. Consider, for example, the third-rank power discussed above. The fact that this power is an emergent high-rank power does not seem to give us any reason to deny that it is a *physical* power.

Kim could close off this objection to his argument by simply stipulating that 'physical' implies non-emergent. If such a stipulation is made, then *Closure* can be rewritten as:

Closure*: If a non-emergent event has a cause, it has a non-emergent cause. And if a non-emergent event has an explanation, it has a non-emergent explanation.

Figure 10.3. Kim's argument from closure

Without the above stipulation about the meaning of 'physical', *Closure* may seem perfectly acceptable to emergentists. *Closure**, on the other hand, is simply the statement that emergent properties cannot have novel causal powers to affect non-emergent properties, and thus seems to beg the question against emergentism.

So is there any independent reason to believe *Closure**? Kim says that if we deny closure, then 'we are asserting that an ideally complete physical theory will not be able to give an account of all physical phenomena' (2006a: 200), and this does seem like an unfortunate thing to have to say. But if we bear in mind that Kim is simply stipulating that emergent properties are not physical, then what the denial of *Closure** really amounts to is the claim that an ideally complete theory of low-level properties and their interactions will not be able to account for all physical phenomena. But this denial is precisely the point of emergentism and so it would beg the question to rule out such a possibility from the start.

As noted above, high-rank powers would count as strongly emergent in the sense defined by Wilson. Like Kim, Wilson takes *Closure** to be 'the core physicalist claim' (2015, 358), and she therefore asserts that strong emergence is incompatible with physicalism. Unlike Kim, however, Wilson does not take this incompatibility as an argument against strong emergence, seeing it instead as simply a character-istic feature of strong emergence. I think, perhaps, that things will be clearer if we do not define 'physicalism' in terms of *Closure**, but instead take *Closure** to be definitive of 'microphysicalism'.

A separate argument for *Closure** is based on a worry that the laws governing low-level properties already give a complete account of the behaviour of low-level entities, so that adding a new high-level property will either make no difference or the high-level laws will conflict with low-level laws. Robert Van Gulick, for example, says the following in his discussion of 'radical' emergence:

> If wholes or systems could have powers that were radically emergent from the powers of their parts in the sense that those system-level powers were not determined by the laws governing the powers of their parts, then that would seem to imply the existence of powers that could override or violate the laws governing the powers of the parts...It is in this respect that radically emergent powers would pose such a direct challenge to physicalism, since they would threaten the view of the physical world as a closed causal system.   (Van Gulick, 2001, 18–19)

But if emergent properties are high-rank powers, then this view is simply mistaken. Adding new influences to a system can make a difference to the system's behaviour, but we can nonetheless add as many new high-level laws of influence as we want without conflicting with the low-level laws. Laws of influence do not purport to describe the ultimate behaviour of a system; they only describe the influences at play. These influences must then be combined to determine the resulting behaviour. Thus, laws of influence cannot come into conflict with each other. This, for example, is why the laws of gravitation do not conflict with the laws of electromagnetism. Adding new laws of influence, emergent or not, just adds new influences that must be combined. I conclude that there is no independent reason to assume *Closure**, and so the argument from closure simply begs the question against emergentism.

10.4.2 The Direct Argument

Kim's second line of argument avoids both *Closure* and *Closure**, and attempts to derive the causal inefficacy of emergent properties directly from the fact that they supervene on non-emergent properties (Kim 1999; 2006b). Kim asks us to consider a case in which some emergent property M, with lower-level supervenience base P, causes some lower-level physical state P^*, as in Figure 10.4.

Now, says Kim, suppose that causation is nomological sufficiency. Then M is, by supposition, nomologically sufficient for P^*. But note that since M nomologically supervenes on P, P is nomologically sufficient for M. Now, if P is nomologically sufficient for M and M is nomologically sufficient for P^*, then, by transitivity, P is nomologically sufficient for P^*. Hence P is a cause of P^*. We get the same conclusion, says Kim, if we suppose instead that causation is counterfactual dependence.

Furthermore, Kim argues, the situation cannot be one where there is a causal chain from P to P^* via M, since the relation of emergence between P and M is not a causal one.[3] As with the argument from closure, then, we must either conclude that P is the cause of P^* and M is not, or conclude that P^* is overdetermined by P and M. Kim argues that systematic overdetermination of this kind can be ruled out on principle, and so he concludes that M must be causally inert. But even if we are willing to countenance such overdetermination, Kim's argument would have us conclude that M is at best redundant and so does not bring with it any new causal powers—contra the claims of emergentism.

I have argued elsewhere (Corry, 2013) that Kim's argument fails due to an equivocation over the concept of causation at play. However, if emergent properties are high-rank powers, then the argument fails in an even more

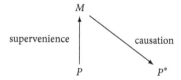

Figure 10.4. A case of downward causation

---

[3] This is supposed to be part of the concept of emergence. O'Connor and Wong 2005 dispute this, however.

straightforward manner. For the whole set-up depicted in Figure 10.4 is misleadingly inappropriate.

Although Kim intends his argument to apply to the possibility of emergent properties in general, the case that he has in mind is the possibility that mental states are emergent. In this case P and P^* represent states of some physical system (patterns of activation in a network of neurons, say), while M is supposed to represent some mental state (believing that it is time to get out of bed, for example). But remember that emergentism is not a form of substance dualism. When the emergentist says that M is an emergent mental state, she is not saying that M represents the state of some non-physical mental system. Rather, she is saying that M is an emergent property of the very same physical system whose non-emergent properties are represented by P. Thus, Kim's argument must conclude that the emergent properties of a system cannot have effects over and above those of the non-emergent properties of that system.

Consider, then, how we might try to apply Kim's objection to emergent properties conceived as high-rank powers. In particular, let us focus on the third-rank power described in section 10.3.1 above. Let P represent the positions and charges of the three particles depicted in Figure 10.1. That is to say, P represents the state of all the properties that the particles can have in isolation. The emergent property in this case is the imagined third-rank power and so M represents the state of this third-rank power. But what could we mean by 'the state of a causal power'? If the causal power were of the kind described in typical counterfactual terms, so that it is either active or not, then M would simply represent whether or not the power was active. More generally, if the power can be active in many different ways, as allowed in my functional description of powers, then M must represent the particular way in which the power is active. Now, causal powers, as I understand them, are dispositions to exert causal influence, so the different ways of being active are simply exertions of different influences. Thus, M will represent the particular influence that is exerted by our emergent power. In the case of our third-rank power, the influence it exerts is simply a force that acts on particle a, and so M will represent the exertion of this force. Note that the force that is exerted by our third-rank power is determined by the charges and positions of the three particles. Since charge and position are properties that an object can have in isolation, it follows that M is determined by, and so supervenient upon, P, as Kim requires.

Finally, if particle a accelerates, then its position will change and the three-particle system will no longer be in state P but some other state P^*, in which case M seems to be at least a partial cause of state P^*. Putting all of this together, the third-rank power described above would seem to give rise to a case of downward causation, as depicted in Figure 10.4. Thus, if Kim's argument is cogent, we must conclude that M is either causally inert or redundant. Since M represents the activity of our posited third-rank power, it would seem to follow that our third-rank power cannot be a non redundant cause of the change to P^*, and so is not really a basic causal power at all.

To see that there is a problem with applying Kim's argument to third-rank powers in this way, note that nothing in the reasoning above was specific to powers of rank three; the reasoning would apply just as well to second-rank powers. Thus, if Kim's argument rules out the possibility of genuine third-rank powers, as above, then it will also rule out the possibility of genuine second-rank powers. Recall, for example, that the gravitational force that one object exerts on another is nomologically determined by, and so supervenient upon, the masses and positions of both objects, so gravitational powers are second-rank. If Kim is right that such supervenient properties cannot be non-redundant causes, then it would follow that an object's gravitational power cannot be a non-redundant cause of a change in the acceleration of another object. Surely, this must count as a reductio of such an application of Kim's argument.

So what has gone wrong in our application of Kim's argument in the case of the second-rank power of gravitational attraction? I believe that the problem comes from confusing a power with the influence that it manifests. The gravitational power that one object has to attract another is, or is represented by, its mass, and the mass of an object is a property that it can have in isolation, even though its gravitational power may only be manifested in a system of at least two massive objects. Thus, the mass of each object is included in the specification of the state P. Since the gravitational powers of the objects involved are partly constitutive of the state P, there can be no sense in which these gravitational powers compete with P as possible causes of P^*.

But now consider M in our example of gravitational attraction. M, here, is not the gravitational power of either object, but a force exerted by these powers in the situation described by P. This force supervenes on P, but does not supervene on the state of any smaller constituents of P, and so, unlike gravitational power, this force is not a constituent of P. Nor could M be identical to P, since influences, and therefore forces in particular, are entities that are distinct from the powers that manifest them. So M, it seems, is a legitimate competitor with P for the title of the cause of P^*. Since M supervenes on P, therefore, Kim's argument implies that M is either not a cause of P^* or is a redundant cause. What Kim's argument tells us is not that gravitational powers are not genuine causes, but rather that gravitation forces (and, by extension, forces in general) are not causes.

The same thing can be said in the case of our third-rank power and, indeed, high-rank powers in general. Since the existence of a basic power is not implied by the existence of any other basic powers, a complete description, P, of the physical state of a system must include all basic causal powers of the constituents of the system, regardless of what rank those powers may be. It is not these powers that are represented by M in Kim's diagram, but rather a *manifestation* of these powers, an influence. Kim's argument, therefore, does not imply any difficulty in regarding powers (of any rank) as causes. Rather, Kim's argument suggests that we will run into difficulties if we regard influences as causes (and this will be true regardless of the rank of the power that manifested the influence).

Are we still left with a problem? Should we be worried that Kim's argument tells us that we will get into trouble if we regard influences as causes? I do not think so. I have already argued in Chapter 8 that influences are not causes. Rather, I suggest, influences are that in the world that ground our claims about cause and effect. Oversimplifying, we can think of influences as physical connections that typically connect causes to effects. Thus, a gravitational force does not cause an acceleration. Rather, the force is the *causing* of that acceleration by the arrangement of gravitational powers. Seen in this light, Figure 10.4 takes on a very different aspect: P and M do not compete as possible causes of P^*. Rather, M is the manifestation of P's power to cause P^*. If this causing involves high-rank powers, then we have a case of emergent causation, but apart from that the rank of the powers makes no difference.

As discussed in Chapter 8, there does not seem to be a single concept of causation, and so it is possible that someone could take a different view of causation from that outlined in the paragraph above, and insist that forces, and influences more generally, do indeed cause changes. But even if we take this view, I do not think Kim's argument represents a problem. For if forces can cause accelerations, then it would seem just as legitimate to say that causal powers arranged certain ways can cause forces to come into existence. In this case P is a cause of M which is a cause of P^*. But, if M is an intermediary in a causal chain from P to P^*, then, once again, M and P are not in competition as causes of P^*.

Thus, neither the argument from causal closure nor the direct argument gives us a reason to object to high-rank powers, and so does not give us reason to object to ontological emergence in this form. Before moving on, however, I wish to consider one more argument that Kim presents against the possibility of downwards causation. Unlike his other arguments, this last argument is directed only against a specific kind of downward causation, but, even in this case, we will see that the argument does not work against downward causation from high-rank causal powers.

10.4.3 The Argument against Synchronic Reflexive Downward Causation

Many emergentists have held that some emergent properties have the power to affect the basic physical properties upon which they supervene. Roger Sperry, for example, holds that subjective mental phenomena are emergent properties which supervene on neural activity within the brain. In turn, he says, these emergent properties 'influence and govern the flow of nerve impulse traffic'

(Sperry, 1969, 534). Kim calls causation of this sort *reflexive downward causation*. As a species of downward causation, Kim believes that reflexive downward causation will be ruled out by the arguments we considered above. But Kim believes there is an independent argument against a specific form of reflexive downward causation, namely *synchronic reflexive downward causation* (1999, 28). As the name suggests, synchronic reflexive downward causation occurs when the emergent cause and the lower-level effect are simultaneous. Many philosophers and scientists view simultaneous causation with suspicion and any such suspicion will transfer in full to synchronic reflexive downward causation. However, such suspicion is not universal. Stephen Mumford and Rani Anjum (2011a, 106–29), for example, have argued quite plausibly that *all* causation is simultaneous. Thus, it will be worthwhile to investigate Kim's argument against synchronic reflexive downward causation further.

The basic problem with synchronic reflexive downward causation (apart from the unwieldy name!) is that it seems to involve some kind of circularity of determination. Suppose that a complex whole W has an emergent property M at a given time t. A case of synchronic reflexive downward causation will be a case in which M causes some component, a_i, of W to have property P_i at t. But if M is an emergent property, it will supervene on the micro-state of W, and so the emergence of M at t will depend (amongst other things) on the component a_i having property P_i at t. Thus, we have a situation where W has M at t because a_i has P_i at t, but also a_i has P_i at t because W has M at t.

Kim suggests that the circularity of determination here is unacceptable because it violates the following principle, which he dubs the *causal-power actuality principle*:

Causal-Power Actuality Principle For an object, x, to exercise, at time t, the causal/determinative powers it has in virtue of having property P, x must already possess P at t. When x is caused to acquire P at t, it does not already possess P at t and is not capable of exercising the causal/determinative powers inherent in P.   (Kim, 1999, 29)

If W having M at t causes a_i to have P_i at t, then the causal-power actuality principle implies that a_i could not have already had P_i at t (if a_i already had P_i, Kim says, then W having M at t can have played no role in causing a_i to have P_i at t). But, if a_i did not already have P_i at t, then W could not already have M at t (since the presence of M depends on a_i having P_i). But then, the causal-power actuality principle implies that W having M at t could not have caused anything at time t, and so, in particular, could not have caused a_i to have P_i at t. Thus, the causal-power actuality principle, together with the assumption of synchronic reflexive downward causation, gives rise to a contradiction.

Kim says that he would rather reject the possibility of synchronic reflexive downward causation than reject the plausible causal-power actuality principle. However, there is no conceptual problem with the kind of circularity that Kim is worried about here, and hence no a priori reason to accept the causal-power actuality principle. To see this, consider the well-defined and well-understood theory of Newtonian gravitation. In Newton's theory, gravitational forces act instantaneously at a distance: when I wave my hand, Pluto immediately wiggles (very, very, slightly). Thus, Newton's theory is one that involves synchronic causation. Now, consider a system consisting of two massive particles. The gravitational force exerted by one particle on another at time t is determined by (amongst other things) the distance between the two particles at t. But the distance between the two particles is itself caused to change by (amongst other things) the gravitational forces acting at time t. Thus, Newton's theory furnishes us with a same-level version of the circularity that Kim deems unacceptable. Surely, though, there is nothing conceptually objectionable about this circularity in Newtonian theory. It is true that with the advent of General Relativity we no longer believe that gravitation operates synchronically, but the point I am making here is simply that the question of whether Newton's synchronic theory is true is an empirical question; Newton's theory does not seem to be conceptually problematic. The fact that in Newton's theory forces depend on positions, whilst positions depend on forces, makes the equations of multi-particle systems difficult to solve. However, general solutions to these equations have been found for the two-particle case, and solutions have been found for a number of specific cases involving three and more particles (J. Cartwright, 2013). Thus, there can be no logical inconsistency in this circularity.

The third-rank power discussed above will also exhibit this kind of circularity: the force that one pair of particles exerts on a third is determined by their relative positions, and their relative positions are caused to change by these forces. What is more, since the power involved is a high-rank power, this is a case of downward causation. But the circularity here is no more objectionable than it was in the Newtonian case.

So, what of the causal-power actuality principle? It is easy to see that this principle is not satisfied by Newtonian mechanics. The influence that one particle has on another at time t is determined by the distance, r, between them at t. Now, unless the two particles are in a perfectly circular orbit around each other, the distance between them at any instant before t will not be r. Thus, in almost all cases, the particles must exert the influences they do at time t in virtue of being separated by distance r, even though they do not already possess the property of being this far apart. The fact that the causal-power actuality principle is violated by a theory that is logically consistent and which was held to be the pinnacle of science for over two hundred years suggests that the principle is not obviously true and, at the very least, requires much more argument. We have good reasons for doubting the truth of Newtonian gravitation, but I do not think that its failure to satisfy the causal-power actuality principle is one of them. Without further argument for this principle, Kim's argument against synchronic downward causation fails.

10.5 Assumption 4 and Compositional Emergence

I argued above that the traditional idea of ontological emergence can be understood as referring to cases in which Assumption 1 (Supervenience) holds but Assumption 2 (Limited Rank) does not. Let us turn our attention now to the other assumptions of reductive explanation—what happens if they fail? I pass over Assumption 3 (Changes Determined by Influence), since it is implied by Assumption 4 (Algebra of Composition). Let me note, however, that my ontology of power and influence will not be able to shed much light on a situation in which Assumption 3 fails due to the fact that something other than an influence is involved in producing a change (rather than failing due to a problem with the algebra of composition).

The next assumption to consider, then, is Assumption 4:

Algebra of Composition Let I_{Pa} be the set of influences of type P acting on object a. There exists a two-place operator $\oplus$ such that:

1. The result of influences u_1, u_2, ... and $u_n \in I_{Pa}$ acting together at the same time, and in the absence of any other influences in I_{Pa}, is the outcome that would occur if the influence $(\ldots(u_1 \oplus u_2) \oplus \ldots) \oplus u_n$ acted on its own;

2. $(I_{Pa}, \oplus)$ is a commutative semigroup.

As mentioned in Chapter 5, this assumption is not necessary for the possibility of a reductive explanation. Rather, it is necessary in order to employ the reductive method to discover or justify such an explanation. Thus, a failure of this assumption will not lead to ontological emergence of the kind discussed above. However, such a failure would still mean that there is something about the world that makes it impossible, even in principle, to use the reductive method, and so we might still regard such a failure as involving a type of emergence that is not purely epistemic.

To get a better feel for how composition laws that do not satisfy this assumption can frustrate attempts to apply the reductive method, let us consider a simple example. Suppose that Mary and Tom are the only two candidates in an election for school captain, and the winner will simply be the candidate who gets the most votes. There are three possible outcomes of this election: Mary could be elected school captain, Tom could be elected school captain, or Mary and Tom could get the same number of votes (in which case neither is elected school captain and a new election is held). For concreteness, let us say that the outcome of the election is embodied in an announcement by the returning officer. In such an election, the outcome is determined by the voting behaviour of the individual students at the school, and it seems we should be able to give a reductive explanation of the outcome in terms of these individual behaviours. In such a reductive explanation we can treat the vote that an individual student casts as representing (or possibly even constituting) a causal influence that the student exerts on the outcome. Now, suppose only a single student voted and that she voted for Mary, then the outcome would be that Mary wins. So a vote for Mary represents an influence directed towards the outcome in which Mary wins. Similarly, a vote for Tom represents an influence directed towards an outcome in which Tom wins. If, on the other hand, only one student votes and she ticks neither box, or both boxes, or in some other way votes incorrectly, then this vote will not be counted and neither candidate will win. Hence, an incorrectly completed vote represents an influence directed towards neither candidate winning.

In this situation we have everything we need for a reductive explanation of the outcome of the vote. The outcome is determined by, and so supervenes upon, the behaviours of the individual students in the school. Thus, the supervenience assumption of reductive explanation is satisfied. If we assume that each student's voting preference is fixed, given the background conditions, then we can treat this preference as an invariant causal power to exert an influence on the election.[4] Furthermore, the relevant influences are exerted by individual students upon the returning officer's announcement, and so they are all of rank two. Thus, the assumption of limited rank is satisfied. Finally, we have a rule for composing influences which states that the resultant influence is directed towards Mary winning if more of the component influences are directed towards Mary winning than towards Tom winning, and vice versa, and the resultant is directed towards neither candidate winning in the case where there are the same number of components directed towards Mary as are directed towards Tom.

Despite the fact that a reductive explanation is possible, the reductive *method* for developing such an explanation will fail. Suppose that we are outside observers who do not know the rules of the voting system, and suppose that we wish to use the reductive method to figure out how the system works. The reductive method tells us to study subsystems in relative isolation. Following this advice, for example, we might run an election in which we only allow a single student, Emily, to vote. By observing the outcome of this mini election, we might learn that Emily is disposed to exert an influence directed towards Mary winning. If we conduct similar mini elections for all the students in the school, we can learn what kind of influence each student is disposed to exert. Admittedly, this is a fairly tedious way of studying student dispositions to vote, but good science is often tedious. Fortunately, there are sometimes shortcuts, such as when the disposition we are interested in is correlated with some other, more readily observable property. After testing a number of students, for example, we might find that the students who wear a yellow T-shirt with Mary's picture on it all vote for Mary, those who wear a

---

[4] In a real election, voting preferences are not fixed. People's preferences can be influenced by election campaigns, world events, the weather, possibly even what they ate that morning. A complete reductive explanation might need to take these influences into account. For simplicity, however, I will ignore such complications.

red T-shirt with Tom's picture on it all vote for Tom, and those who wear neither of these T-shirts all vote for neither. In this case we might not bother conducting mini elections with every student. Indeed, our background knowledge might give us enough confidence in these correlations that we do not even bother running the mini elections at all.

So far, then, we have encountered no problem, in principle, with the reductive method for figuring out the relevant causal powers of our basic objects. But observe what happens if we apply the reductive method in an attempt to figure out the composition law for the relevant influences. Once again, the reductive method would have us study simple subsystems. Now, the simplest subsystem that involves composition of influences is a subsystem in which just two relevant influences are at play. We can study such systems by running mini elections, each of which involves two voters whose voting preferences are known. So, for example, if our mini election involves two students who vote for Mary (that is, they both exert an influence directed towards the outcome in which Mary wins), we will observe that Mary wins. Thus, we can conclude that two influences directed towards Mary winning will compose to form an influence directed towards Mary winning. Similarly, we will discover that an influence directed towards Mary winning composed with an influence directed towards neither winning will result in an influence directed towards Mary winning, and so on. If we conduct such mini elections to study all the possible combinations of two influences we will find that they compose as shown in Table 10.1.

If the composition of voting influences forms a commutative semigroup, then we could use this table to calculate the result of any number of influences acting together. So, for example, to calculate the resultant of three influences we could simply compose the first two influences using the table, and then take that resultant and use the table again to compose it with the third. A moment's consideration, however, will show that this procedure will not work in this case, for the composition function defined by our table is not associative. Suppose that we wish to calculate the resultant of one vote for Tom and two votes for Mary. We could begin by composing the vote for Tom with one of the votes for Mary to get an intermediate resultant directed at neither winning. If we then take this intermediate result and compose it with the vote for Mary, the result is an influence directed towards Mary winning. However if we first compose the two votes for Mary, the intermediate resultant is itself a vote for Mary. If we now compose this intermediate resultant with the vote for Tom, the overall result is an influence directed towards neither winning. In other words, $(Tom \oplus Mary) \oplus Mary = Mary$, while $Tom \oplus (Mary \oplus Mary) = Neither$. In general, the result that one gets when composing three or more influences using this table will depend on the order in which one composes them. But since the order does not correspond to anything physical, there is no way to distinguish the different orders, and so no way to use the reductive method to pick one order rather than another.

Table 10.1 Pairwise composition of voting influence

⊕	Mary	Neither	Tom
Mary	Mary	Mary	Neither
Neither	Mary	Neither	Tom
Tom	Neither	Tom	Tom

There is, of course, a very simple composition function for this electoral system: the resultant influence will be directed towards Tom winning if there are more components directed towards Tom than there are directed towards Mary; the resultant will be directed towards Mary winning if there are more components directed towards Mary than there are directed towards Tom; and the resultant will be directed towards neither winning if there are the same number of components directed towards Tom as there are directed towards Mary. The problem is that the correct composition function depends on the statistical properties of the *whole set* of relevant influences, and so it cannot be expressed in a pairwise composition table like Table 10.1. There is a very real sense, therefore, in which the composition function for the whole system of relevant influences is 'greater than the sum of' the composition functions for the parts. In this sense, it seems reasonable to regard the composition of influence in this system as 'emergent'. It is this emergent nature of the composition function which frustrates the reductive method. In general, if Assumption 4 fails in some situation because the composition of relevant influences does not form a commutative semigroup, then the reductive method will fail, and we will have good reason to regard this failure as a case of ontological emergence.

10.6 Assumption 5 and Non-Linearity

I am not the first person to suggest that there may be a form of ontological emergence associated with composition laws. As I mentioned in Chapter 4, Mumford and Anjum (2011a) have developed a 'vector model' of causation that bears many similarities to the metaphysics of power and influence that I am proposing in this book. They do not distinguish powers from the influences that they manifest, and so they speak of the composition of *powers* rather than composition of *influence*, and when they speak of a manifestation, they do not refer to an influence, but rather to the outcome that occurs due to the action of a set of powers—the outcome towards which the resultant influence is directed, as I would say. Mumford and Anjum suggest that a variety of genuine (non-epistemic) emergence may occur if the composition function for powers is truly *non-linear*.

Non-linearity, they say, 'simply involves the output of the system not being proportional to the input' (2011a, 97). As an example of a system that appears to be non-linear they consider Earth's weather system, in which a small input, like the beating of a butterfly's wings, can give rise to a large output, such as a tornado.[5]

This gloss of non-linearity in terms of the inputs and outputs of systems does not make it clear what—if anything—non-linearity has to do with composition. However, Mumford and Anjum do provide another 'more accurate' account of non-linearity that makes the relevance to composition more apparent. A non-linear system, they say, is one that does not obey the principle of additivity, which is that $f(x + y) = f(x) + f(y)$ (2011a, 97). Here the role of composition seems clearer, but there is another problem: Mumford and Anjum do not tell us what x, y, and f are supposed to represent in this equation.

Fortunately, Mumford and Anjum (2011a, 87) provide another account of what they call the 'principle of additive composition', which states the following:

Principle of Additive Composition If $D1$ is the power to $M1$, and $D2$ is the power to $M2$, then $D1$ and $D2$ working together are (nothing more nor less than) the power to $M1 + M2$.

Here, $D1$ and $D2$ are powers, while $M1$ and $M2$ are the characteristic manifestations towards which these powers are directed. In order to reconcile these two accounts of the principle of additivity, then, we need to take x and y in the equation above to represent powers, while f is the function that maps a power to its characteristic manifestation. Then, we can read the equation above as stating that the characteristic manifestation of powers x and y working together is the sum of the manifestations of x and y working individually.

This interpretation brings to light one more problem. The first time the symbol '+' appears in the equation, it represents an operator on the space of powers, but the second time this symbol appears, it represents an operator on the space of manifestations. Since these spaces are quite different, an operator in one space cannot be the same operator that acts in the other, and it is therefore somewhat misleading to use the same symbol for both.

So, let us use the symbol '$\oplus$' to represent the appropriate composition operator in the space of powers, and the symbol '$\star$' to represent composition in the space of manifestations, as I did in Chapter 5. Then, the principle of additive composition is that $f(x \oplus y) = f(x) \star f(y)$. In other words, Mumford and Anjum's 'principle of additive composition' is actually the claim that the function f, which maps a power to its characteristic manifestation, is a homomorphism from the space of powers with the composition operator $\oplus$ to the space of manifestations with the composition operator $\star$.

In Chapter 5, I argued that in order for the reductive method to proceed, the function which maps an influence to its characteristic outcome must be a

---

[5] I say 'seems' because Mumford and Anjum consider the possibility that the apparent non-linearity of the weather might, in fact, be due to a cascade of 'tipping-points'.

homomorphism from the space of influences under the composition operator $\oplus$ to the space of effects under the operator $\star$. It should be clear that Mumford and Anjum's principle of additive composition is just a restatement of this requirement in their language (which does not distinguish power and influence), and so the reductive method will indeed fail if additivity fails.

Now, there are actually two ways in which the principle of additive composition might fail. First, it might fail if composition in the space of powers does not get mapped to the *expected* composition operator in the space of manifestations. That is, we already have some composition operator $\star$ in mind and we find that for some x and y, $f(x \oplus y) \neq f(x) \star f(y)$. In this case there may still exist some operator whose algebra is homomorphic to the algebra of powers, but as this is not the operator we have in mind, the system will appear to resist the reductive method. I am not sure if it would be legitimate to regard this as a case of emergence, since it is not clear that the problem has anything to do with the whole being 'more than' the sum of the parts. But even if we did decide to regard this as a case of emergence, it seems more like a case of epistemic emergence than ontological emergence, since the problem lies with our expectations rather than with the system itself.

The second way in which the principle of additive composition might fail is if there does not exist *any* composition operator, $\star$, such that f is a homomorphism. In this case, as I argued in Chapter 5, the reductive method for justifying, developing, and extending reductive explanation would fail. But, as I also pointed out in that chapter, the only way that there could fail to be an operator that plays the role of $\star$ in the principle of additive composition is if f is not a one-to-one function, which is to say that Assumption 5 (that influences are uniquely characterized by their effects in isolation) fails. Failure of additivity in this sense will lead to problems for the reductive method, but it is not clear that such a failure should be described as a case of *emergence*, since it does not seem to have anything to do with the properties of a whole going beyond the properties of the parts. Rather, it is a case in which two distinct powers (distinct in the sense of occupying different places in the algebra of composition) have the same characteristic manifestation.

Thus, the non-linearity of composition that Mumford and Anjum identify as a possible form of emergence is either merely epistemic emergence or equivalent to a failure of Assumption 5, which, although it would lead to an interesting failure in the reductive method, does not seem to be a case of emergence. I cannot help thinking that Mumford and Anjum really had in mind something more like the failure of Assumption 4 discussed above.

10.7 Conclusion

I have argued that the ontology of power and influence can make sense of the concept of ontological emergence. I have also argued that such ontological emergence cannot be ruled out by the arguments put forward by Jaegwon Kim. The remaining question is whether there is any good empirical evidence for emergence of this kind. There are certainly many systems in which attempts at reductive explanation have failed. Most obvious is the failure of attempts to reductively explain consciousness in terms of the arrangement and behaviour of neurons in the brain. Other examples mentioned by Wilson (2015, 50–1) include failures to reductively explain the so-called 'universal' behaviour of gases near 'critical points' in terms of the systems of atoms or molecules composing them, and failures to reductively explain certain global properties of honeybee hives in terms of the aggregative behaviours of individual bees. As I mentioned above, however, we can never be sure that resistance to reduction is not a case of merely epistemic emergence rather than ontological emergence. Timothy O'Connor (1994) has argued that there is some positive empirical evidence for ontological emergence, citing work by Michael Polanyi (1968) and Roger Sperry (1969), who have hypothesized that ontological emergence is involved in embryonic cells and consciousness, respectively. However, as Mark Bedau (1997) points out, the work of Polanyi and Sperry hardly constitutes a thriving research programme. The papers that O'Connor cites are almost fifty years old, and they have inspired little scientific work since. What is more, it is not obvious that the phenomena that interest Polanyi and Sperry could not be cases of merely epistemic emergence. There have, however, been more recent suggestions of empirical evidence for ontological emergence. In a recent edition of *Philosophica*, for example, Peter Lewis (2017) argues that the quantum mechanical behaviour of entangled particles provides good evidence of the existence of strongly emergent entities; Jonathan Bain (2017) suggests that topological insulators, topological superconductors, and systems which produce the quantum Hall effect support emergentism; and Tom McCleish (2017) argues that numerous biological phenomena—including protein assembly, gene expression, and the topological interaction of DNA and topoisomerase enzymes—exhibit downwards causation in a manner indicative of ontological emergence.

I will not consider these suggestions of empirical evidence for emergence here. Instead, let us consider what could count as positive evidence for ontological emergence. As I noted above, repeated failures of attempts at reductive explanation, especially in cases where the system is not too complex, and the components in isolation are well understood, may constitute evidence for ontological emergence. But such evidence would still have a negative aspect—it is still evidence based on the failure of reduction. The best positive evidence for a previously unknown property or entity comes when this entity is an essential part of a successful theory—a theory that is able to explain and predict many phenomena. But if all we can say about emergent properties is that they are properties of complex systems which somehow go beyond the properties of the parts, then the best we will be able to do is explain and predict a certain pattern of failure. We will have better reason to believe in emergent properties if they play a role in making positive predictions. This kind of positive evidence, however, requires a positive characterization of emergence, so, given the rather vague accounts of emergence that currently prevail, it is no wonder that emergence has not played a positive role in scientific theory. The ontology of power and influence, on the other hand, provides such a positive characterization, and shows that emergent powers and emergent composition laws are not that different from the powers and laws that we already accept. It is now a purely empirical matter whether these emergent powers and composition laws will play a useful role in understanding the world.

11

Influentialism

A New Type of Moral Theory?

In his book, *Powers*, George Molnar states that 'the compelling reasons for accepting the theory of powers outlined are in the work that the concept can do in one's metaphysics' (2003, 186). I, too, have been concerned with the work that a metaphysical concept can do for us. Throughout this book, my strategy has not concentrated on arguing the faults of metaphysical theories that differ from mine. Rather, my strategy has been to provide a positive argument for a metaphysics of power and influence by showing the work that this metaphysical picture can do for us. In particular, I have argued that a metaphysics of power and influence can help us better understand the reductive method of explanation, the nature of causation, causal models, laws of nature, and the possibility of emergence. In this final short chapter I wish to extend the investigation beyond metaphysics by suggesting that the ontology of influence may be helpful in the field of normative ethics. In particular, I will argue that the ontology of causal influence opens up the possibility of a novel category of normative ethical theory which I will call *influentialism*. Influentialism stands in contrast to the traditional categories of consequentialism, deontology, and virtue ethics. My aim is not to argue that influentialism is preferable to these traditional categories. I simply hope to put the theory on the table for consideration. But I will go so far as to argue that influentialism has some promising features that make it worthy of consideration. In particular, influentialism seems to occupy a middle ground between consequentialism and deontology and is able to combine seemingly incompatible intuitions from these two categories.

In order to help motivate the new theory, I will begin with a discussion of some problem cases for consequentialism in Section 11.1. The point of these cases is to highlight a class of situations in which some intuitions may diverge from the dictates of consequentialism. In Section 11.2 I outline the theory of influentialism and show how it can avoid the problem cases. Finally, in Section 11.3, I discuss influentialism's relation to other categories of normative ethics.

11.1 Problem Cases for Consequentialism

11.1.1 Joining the Lynch Mob

Imagine you have recently moved to a small town. You are out for a stroll one afternoon when you witness a large crowd gathered in the town square. When you approach to see what the fuss is about, you discover the crowd is actually a lynch mob, and they are about to hang someone, ostensibly for some petty misdemeanour. With horror in your heart, you realize that there is nothing you can do to stop the lynching—the mob is too large for you to stop them by force, they are clearly beyond reason at this stage, and you can see the local policeman in the mob, so calling the authorities is not going to help. But as you struggle to decide whether you should raise your voice against the lynching anyway, or simply run away and cry, things get worse. One of your new neighbours recognizes you and invites you to put the rope around the victim's neck. The offer is obviously made as a test of where you stand, and you can be sure that if you refuse, then life will not be easy in your new town. What should you do?

Consequentialist theories assert that the morality of an action is to be judged according to the value of the consequences of that action. So, given that the lynching will happen whether you participate or not, the death of a relative innocent should play no part in any consequentialist deliberation about whether to participate or not. In this case, your decision whether or not to participate is likely to have the largest effect on you, and so it seems that if the bad consequences for you if you don't participate (being ostracized from the community, say) outweigh the bad consequences for you if you do, then you should participate in the lynching.

Of course, a consequentialist might point out that participating in the lynching may have terrible psychological consequences for you—you may be racked with guilt; you may suffer nightmares; participation may even rupture your sense of who you are. In this case, consequentialism could tell you not to participate. But we can modify the scenario slightly: what if, deep down, you wanted to participate? Perhaps the victim is a thoroughly unlikable person, and perhaps you are a dyed-in-the-wool consequentialist, so that you will feel no guilt in participating if that is what consequentialism dictates. In short, suppose that you will suffer no bad consequences if you participate—would this make it right to join in? Would wanting to be part of the lynch mob really make it permissible—nay, obligatory—to take part?

This thought experiment is very similar in structure to two thought experiments put forward by Bernard Williams (1973) in criticism of consequentialism. Williams did not argue that in cases like this consequentialism gives the wrong answer regarding what one should do, rather, he worried that consequentialist thinking leaves something important out of the picture. As he puts it, consequentialism 'cuts out a kind of consideration which for some others makes a difference to what they feel about such cases: a consideration involving the idea, as we might first and very simply put it, that each of us is specially responsible for what he does, rather than for what other people do' (1973, 99). Williams is concerned that, by focusing solely on states of the world, consequentialism ignores the role that agents play in contributing to those states. Put simply, the problem with joining the lynch mob is that in doing so you would be participating in an act that is morally reprehensible.

Fortunately, the lynch-mob scenario is rather extreme, and most of us are never likely to find ourselves in such a situation. However, I believe that most of us do find ourselves in situations with a very similar structure on a regular basis. The next example makes this clear.

11.1.2 The Child Sweatshop

Suppose that you are shopping for a new football for your child, and you have a choice between Brand X and Brand Y. The brands are both large multinational companies, and the two kinds of ball are equally good, but there is a big difference in how they are made. Brand Y balls are made in a pleasant factory by adults who work an eight-hour day and earn a living wage. Brand X balls, on the other hand, are made in terrible conditions by children who work a twelve-hour day and are not paid enough to keep them out of poverty. As a result, Brand X balls are slightly cheaper than Brand Y. Which brand should you buy?

For the purposes of argument, let us put aside any discussion over the morality of child labour, and simply accept that the child sweatshops that produce Brand X balls are morally reprehensible. If we are concerned with the consequences of our actions, then the first thing to note is that multinational sports-equipment companies measure their sales in units of hundreds or thousands. As a result, their decisions as to where to manufacture the balls, or how many balls to manufacture, will almost certainly be completely insensitive to whether or not you buy one of their balls. When deciding whether or not to purchase Brand X, then, you are in a similar situation to that described in the lynch-mob example. To buy Brand X is to participate, to some extent, in the ongoing exploitation of children, but on the other hand, deciding not to buy Brand X will make absolutely no difference to the suffering of those children.

As with the lynch-mob case, it seems that consequentialist theories may tell us that the right thing to do is participate in the moral outrage. After all, you could use the money you save to buy yourself a treat, and thereby increase the total happiness in the world.

11.1.3 Voting

One more example: if voting is not compulsory, then at election time you are faced with a decision whether or not to make the effort to vote. Suppose then that it is voting day in a national election, and that there is one candidate, Smith, who is clearly better than the alternatives—perhaps she is the only candidate who could avert a looming war, or be the catalyst who brings about international action on climate change. In short, the election of Smith would be a highly valuable outcome if you could act so as to make it happen. But the reality is that in a national election you will be part of a large electorate, and your vote will almost certainly make no difference at all to the outcome of the election. Indeed, the probability of the election being decided by one vote will be so low that the possibility can surely be discounted even in a calculation of expected utility.

So, should you vote? Some of us might think that there is a duty to vote, but consequentialist theories will disagree. Given that the outcome of the election will almost certainly be the same regardless of whether you vote (and regardless of who you vote for), the outcome should not be included in a consequentialist evaluation of voting. But if we rule the outcome of the election out of consideration, then there seems very little value in voting for the clearly better candidate, or in voting at all. Perhaps you could raise the total happiness in the world more by sleeping in, or spending the day at the beach, rather than voting. Consequentialism, therefore, seems to tell us that we have no moral obligation to vote. In fact, in many normal circumstances, consequentialism suggests that we may be morally obliged not to vote.

As I stated in the introduction, none of these cases is intended as a knock-down argument against consequentialism. Consequentialists could argue that if we take all of the consequences into account, then their preferred theory does not draw the conclusions I have suggested above. Alternatively, consequentialists could bite the bullet and insist that any intuitions one may have that differ from the dictates of consequentialism in these cases are simply mistaken. The purpose of these cases is, rather, to open a discussion about a certain type of situation. I will then suggest that the influentialist theory to come may have something interesting to say in these cases.

The lynch mob, sweatshop, and voting cases discussed above share a certain causal structure. They are all cases in which you are deciding whether or not to participate in bringing about some outcome (bad in the first two cases, good in the third) when your participation will, in fact, make no relevant difference to the outcome. Because your decision makes no relevant difference to the outcome in question, consequentialism will simply ignore that outcome in its evaluation of your act. Even those with consequentialist leanings might be worried by these situations, however, because although your actions do not make a difference to the outcome, they nonetheless seem to contribute to that outcome. In the sweatshop case, for example, you are contributing just as much as any other private consumer, and ongoing private consumption could well be the main cause of the ongoing sweatshop. As E. F. Schumacher once said:

> We must do what we conceive to be the right thing and not bother our heads or burden our souls with whether we will be successful. Because if we don't do the right thing, we will be doing the wrong thing and we will just be a part of the disease and not a part of the cure. (Schumacher, 1979, 100)

11.2 Influentialism

11.2.1 The Basic Idea

The point of this chapter is that the metaphysics of causal influence opens the possibility of a new class of normative ethical theories which seem better equipped than consequentialism to handle the kinds of situation described above. The basic idea is that, rather than evaluating an act by focusing on the outcomes it leads to, we can focus on the causal influences that are contributed by the act, and so I dub this class of theories *influentialism*.

Consider again the lynch-mob and sweatshop cases. Consequentialism ignored the large harm being done in these situations, because your actions could not prevent this harm. But your choice does affect the kind of influence that you contribute to the situation. If you participate, then you are contributing an influence directed towards a bad outcome. If you do not join in, you will not contribute this negative influence and you may contribute an influence (however futile) directed towards a good outcome. Thus, if actions are evaluated by the influences they contribute rather than by the outcomes they produce, then we may be able to insist that it is wrong to participate in these situations. Influentialism, unlike consequentialism, seems able to capture the intuition that there is something wrong with contributing to a bad outcome, even if one is powerless to prevent it.

Similarly, in the voting case, if you vote (and vote well), then you contribute an influence directed towards a positive outcome. If you fail to vote, however, you forgo the opportunity to contribute such an influence. Thus, influentialism may tell us that voting is the right thing to do even if your actions cannot change the outcome of the election.

As discussed in Chapter 10, electoral systems can combine votes in a threshold manner that may frustrate attempts to apply the reductive method. Nonetheless, reductive explanations are still possible, and, more importantly, we can still speak of causal influences being exerted by the casting of votes. So, even though your vote may make literally no difference to the outcome of the election, a realist about causal influences can hold that by voting you are contributing an influence which is directed towards a certain outcome.

11.2.2 Working Out the Details

Influentialism is the view that we should evaluate an action based on the influences that the action contributes. To produce a theory with any content, many details still need to be filled in. One might wonder, for example, what makes for a good influence, or how we should evaluate actions that contribute multiple influences. Different ways of answering these questions will lead to different theories, which is why I class influentialism as a high-level theory on a par with consequentialism, deontology, or virtue ethics. I do not wish to argue for any particular version of influentialism here, but I will briefly point to some ways that these questions might be answered in order to show how particular theories might go.

The obvious way to evaluate an influence is in terms of the outcome towards which it points. At this stage one will need to supplement the theory with a meta-ethical theory about what counts as a good outcome. In this respect influentialism is in the same boat as consequentialism. Perhaps the best outcome is one that maximizes happiness, or maybe there are other intrinsic values that should be promoted, or perhaps a good outcome is simply one that is loved by the gods.

Unlike consequentialism, however, influentialism has a second dimension that could play a role in the evaluation of influences, for an influentialist theory might plausibly take into account the strength of the influence. Perhaps a strong influence directed towards a moderately good outcome will be preferable to a weak influence directed towards a better outcome. Indeed if we have a way to assign quantitative measures to various outcomes (as some utilitarians attempt to do), and to the strengths of influences (as we have in the case of forces), then we could assign quantitative measures to an influence in proportion to the strength of the influence multiplied by the value of the outcome towards which it points.

What do we do about the fact that a single act may contribute many influences? One option is to take inspiration from utilitarianism and insist that one must act so as to maximize the total value of all the influences one's actions contribute. Such a theory would require both a quantitative account of influence value and an account of how such values are to be added. A less demanding theory (both in terms of the theoretical resources required and in terms of what it demands of us as moral agents) might simply insist that the total strength of positive influences outweigh the total strength of negative influences. Such a theory would not require quantitative normative evaluations of influences, and would simply refer to the method of composing influences that is supplied by the metaphysics of causal influence.

Another possible approach to the problem of multiple influences would be to take inspiration from some deontological theories and insist that only the value of the *intended* influence be taken into account when evaluating an action. Just as deontological theories that focus on intentions give rise to the doctrine of double effect (see McIntyre, 2014), an influentialist theory that focuses on intentions may give rise to a 'doctrine of double influence' according to which an action may be morally right even if it is known to contribute very negative influences, so long as the intention is to contribute a positive influence.

11.3 Relations to Other Normative Theories

None of the cases discussed above is particularly problematic for deontological or virtue-ethical theories, since these theories are not primarily concerned with consequences. A deontologist, for example, might tell us that we should not participate in the lynching because it would violate the rule that one should not participate in the harming of an innocent. A virtue ethicist might insist that such participation would not express the virtue of compassion. The point of putting influentialism on the table, then, is not that influentialism is the only theory that can deal with these problem cases. Rather, the point is that influentialism seems to combine many of the attractive features of consequentialism and deontology.

Influentialism is clearly quite similar to act consequentialism in so far as both share a concern for the outcomes towards which an act is directed. Indeed, in cases over which you have a large influence (in particular, cases in which you can contribute a strong positive influence that is likely to succeed in bringing about the end it is directed towards), influentialism and act consequentialism are likely to give the same verdict on how you should act. As we saw in the examples above, however, influentialism and act consequentialism may come apart in cases where you cannot contribute a significant influence.

In fact, influentialism seems to occupy a middle ground somewhere between consequentialist and deontological theories. Like consequentialist theories, influentialism assesses actions based on consequences. Like deontological theories, influentialism demands that we 'do the right thing' regardless of the consequences. This apparently contradictory position is possible because influentialism decides which actions are right based on the consequences towards which influences are directed, rather than the consequences that would actually occur (or even the consequences that are thought likely to occur).

Another normative theory that attempts to occupy the middle ground between consequentialism and deontology is rule consequentialism (Brandt, 1959; Harsanyi, 1977; Hooker, 2000). Like deontological theories, rule consequentialism demands that we act in accordance with certain rules. However, rule consequentialism states that the correct rules to follow are those that would have the best consequences if accepted by everyone.

It may seem that rule consequentialism can accommodate the intuition that you should not participate in a lynching or buy the products of child labour. For if everyone acted in accordance with this intuition, then there would be no lynchings and no market for the products of child labour. But things are not so simple. Given the unfortunate fact that people do not universally act in accordance with such intuitions, you could find yourself in a situation like those described above, where you cannot prevent some bad thing happening, and you will be better off, in fact, if you participate. Rather than following the rule 'Do not participate in lynchings', then, a better outcome might be obtained if everyone followed the rule 'Do not participate in lynchings unless you have little chance of preventing the lynching and you will be better off if you participate'.

This problem with rule consequentialism is an instance of the old complaint that rule consequentialism collapses into act consequentialism (Lyons, 1965, 133).

Consider a general rule like 'Don't lie'. Suppose there is some situation X in which it is better overall to break this rule. Then rule consequentialism seems to tell us to accept instead the modified rule 'Don't lie except in situations like X'. Since this point generalizes, it seems that rule consequentialism will ultimately endorse only rules that are so specific that they apply just to single acts, and hence rule consequentialism will not differ in practice from act consequentialism (though see Hooker, 2000, 92–9) for a defence of rule consequentialism against this objection).

Because influentialism focuses on acts from the outset, it is not open to the threat of collapse that faces rule consequentialism. Nonetheless, influentialism is in a position to capture one motivation behind rule consequentialism—namely that we all need to do our bit to contribute to the best outcome, even if our individual contributions make little difference to the outcome.

11.4 Conclusion

Although I have presented some problem cases for consequentialism, and have argued that influentialism may have the resources to handle these cases, my aim is not to argue that influentialism is preferable to consequentialism. This chapter is more modest. My aim is simply to point to a new region in the logical space of normative ethical theories, and argue that this region accords with some intuitions in a way that makes it worth investigating further.

A less direct, though more important aim of this chapter is to show that the metaphysics of power and influence may have utility beyond the traditional concerns of metaphysics. In the first five chapters of this book I developed the metaphysics of power and influence, and argued that it is presupposed by attempts at reductive explanation, and by the reductive method of developing, justifying, and extending such explanations. In the next five chapters I argued that the ontology of power and influence can help us make sense of the metaphysics of macroscopic powers, laws of nature, causation, causal models, and emergence. Finally, in this chapter I have suggested that the ontology of power and influence might open interesting possibilities in normative ethics. This is a broad range of applications, but I cannot help feeling that I have merely scratched the surface. It seems to me that an ontology of power and influence may help us understand how we individuate objects, how we understand field theories in physics, and maybe even how we assign moral responsibility.[1] Unfortunately, exploration of these topics will have to wait for another day. For the moment, however, I hope I have done enough to establish that the ontology of power and influence should be taken seriously, and is worthy of further investigation.

---

[1] Nicholas Koziolek (2018) has recently argued that something like my ontology of power and influence can also help us make sense of what it is to believe something.

Bibliography

Achinstein, Peter (1975). 'Causation, Transparency, and Emphasis'. *Canadian Journal of Philosophy*, 5(1): 1–23. http://www.jstor.org/stable/40230552.

Ahmed, Arif (2007). 'Agency and Causation'. In Huw Price and Richard Corry (editors), *Causation, Physics, and the Constitution of Reality: Russell's Republic Revisited*, pp. 120–55. Oxford University Press, Oxford.

Alexander, Samuel (1920). *Space, Time, and Deity*. Macmillan, London.

Anjum, Rani Lill and Mumford, Stephen (2011). 'Dispositional Modality'. In C. F. Gethmann (editor), *Lebenswelt und Wissenschaft*, volume 2 of *Deutsches Jahrbuch für Philosophie*, pp. 468–82. Meiner Verlag, Hamburg.

Anscombe, G. E. M. (1971). *Causality and Determination: An Inaugural Lecture*. Cambridge: Cambridge University Press, Cambridge.

Armstrong, David (1983). *What is a Law of Nature?* Cambridge University Press, Cambridge.

Armstrong, David (1997). *A World of States of Affairs*. Cambridge University Press, Cambridge.

Baetu, Tudor M. (2015). 'The Completeness of Mechanistic Explanations'. *Philosophy of Science*, 82(5): 775–86. doi:10.1086/683279.

Bain, Alexander (1870). *Logic*, volume III. Longmans, Green, Reader & Dyer, London.

Bain, Jonathan (2017). 'Topological Order and Emergence'. *Philosophica*, 91(2): 77–112. http://www.philosophica.ugent.be/fulltexts/92-4.pdf.

Baumgartner, Michael (2008). 'Regularity Theories Reassessed'. *Philosophia*, 36(3): 327–54. doi:10.1007/s11406-007-9114-4.

Baysan, Umut and Wilson, Jessica (2017). 'Must Strong Emergence Collapse?' *Philosophica*, 91(1): 49–104. http://www.philosophica.ugent.be/fulltexts/91-3.pdf.

Bechtel, William and Abrahamsen, Adele (2005). 'Explanation: A Mechanist Alternative'. *Studies in History and Philosophy of Science Part C: Studies in History and Philosophy of Biological and Biomedical Sciences*, 36(2): 421–41. doi:10.1016/j.shpsc.2005.03.010.

Bechtel, William and Richardson, Robert C. (2010). *Discovering Complexity: Decomposition and Localization as Strategies in Scientific Research*. MIT Press, Cambridge, MA. https://mitpress.mit.edu/books/discovering-complexity.

Beckermann, Ansgar, Flohr, Hans, and Kim, Jaegwon (editors) (1992). *Emergence or Reduction? Essays on the Prospects of Nonreductive Physicalism*. De Gruyter, Berlin.

Bedau, Mark A. (1997). 'Weak Emergence'. *Noûs*, 31: 375–99. doi:10.1111/ 0029-4624.31. s11.17.

Bender, Andrea and Beller, Sieghard (2011). 'Causal Asymmetry across Cultures: Assigning Causal Roles in Symmetric Physical Settings'. *Cultural Psychology*, 2: 231. doi:10.3389/fpsyg.2011.00231.

Benthem, Johan (1991). *The Logic of Time: A Model-Theoretic Investigation into the Varieties of Temporal Ontology and Temporal Discourse*. Springer, Dordrecht, second edition. http://public.eblib.com/choice/publicfullrecord.aspx?p=3106158.

Bickle, John (1998). *Psychoneural Reduction: The New Wave*. MIT Press, Cambridge, MA.

Bird, Alexander (2004). 'Antidotes All the Way Down?' *Theoria*, 19(3): 259–69.

Bird, Alexander (2007). *Nature's Metaphysics: Laws and Properties*. Oxford University Press.

Bird, Alexander (2016). 'Overpowering: How the Powers Ontology Has Overreached Itself'. *Mind*, 125(498): 341–83. doi:10.1093/mind/fzv207.

Bostock, David (1979). *Logic and Arithmetic, Vol. 2: Rational and Irrational Numbers*. Clarenden Press, Oxford.

Bostock, Simon (2001). 'The Necessity of Natural Laws'. PhD thesis, University of Sheffield.

Brandt, Richard B. (1959). *Ethical Theory*. Prentice-Hall, Englewood Cliffs, NJ.

Brigandt, Ingo and Love, Alan (2017). 'Reductionism in Biology'. *The Stanford Encyclopedia of Philosophy (Spring 2017 Edition)*. https://plato.stanford.edu/archives/spr2017/entries/reduction-biology/.

Broad, C. D. (1925). *The Mind and its Place in Nature*. Kegan Paul, Trench, Trübner & Co, London.

Carroll, John W. (1994). *Laws of Nature*. Cambridge University Press, Cambridge and New York.

Cartwright, Jon (2013). 'Physicists Discover a Whopping 13 New Solutions to Three-Body Problem'. *Science*, (March 8). http://www.sciencemag.org/news/2013/03/physicists-discover-whopping-13-new-solutions-three-body-problem.

Cartwright, Nancy (1980). 'Do the Laws of Physics State the Facts?' *Pacific Philosophical Quarterly*, 61(1–2): 75–84. doi:10.1111/j.1468-0114. 1980.tb00005.x.

Cartwright, Nancy (1983). *How the Laws of Physics Lie*. Oxford University Press, Oxford.

Cartwright, Nancy (1989). *Nature's Capacities and their Measurement*. Oxford University Press.

Cartwright, Nancy (1999). *The Dappled World: A Study of the Boundaries of Science*. Cambridge University Press, Cambridge.

Cartwright, Nancy (2007). *Hunting Causes and Using Them: Approaches in Philosophy and Economics*. Cambridge University Press.

Cartwright, Nancy (2009). 'Causal Laws, Policy Predictions, and the Need for Genuine Powers'. In Toby Handfield (editor), *Dispositions and Causes*, Mind Association Occasional Series, pp. 127–57. Oxford University Press, Oxford.

Chalmers, David J. (2006). 'Strong and Weak Emergence'. In Paul Davies and Philip Clayton (editors), *The Re-Emergence of Emergence*, pp. 244–54. Oxford University Press.

Choi, Incheol, Nisbett, Richard E., and Norenzayan, Ara (1999). 'Causal Attribution across Cultures: Variation and Universality'. *Psychological Bulletin*, 125(1): 47–63. doi:10.1037/0033-2909.125.1.47.

Choi, Sungho (2005). 'Dispositions and Mimickers'. *Philosophical Studies: An International Journal for Philosophy in the Analytic Tradition*, 122(2): 183–8. https://www.jstor.org/stable/4321552.

Choi, Sungho (2006). 'The Simple vs. Reformed Conditional Analysis of Dispositions'. *Synthese*, 148(2): 369–79. doi:10.1007/s11229-004-6229-z.

Choi, Sungho (2008). 'Dispositional Properties and Counterfactual Conditionals'. *Mind*, 117(468): 795–841. http://www.jstor.org/stable/20532697.

Churchland, Patricia Smith (1986). *Neurophilosophy: Toward a Unified Science of the Mind-Brain*. Computational Models of Cognition and Perception. MIT Press, Cambridge, MA.

Clayton, Philip and Davies, Paul (editors) (2006). *The Re-Emergence of Emergence*. Oxford University Press, Oxford.

Contessa, Gabriele (2014). 'Only Powers Can Confer Dispositions'. *The Philosophical Quarterly*, 65(259): 160–76. doi:10.1093/pq/pqu076.

Corradini, Antonella and O'Connor, Timothy (editors) (2010). *Emergence in Science and Philosophy*. Number 6 in Routledge Studies in the Philosophy of Science. Routledge, New York.

Corry, Ben (2016). 'Computer Simulation of Ion Channels'. In Carmen Domene (editor), *Computational Biophysics of Membrane Proteins*, pp. 161–96. Royal Society of Chemistry, Cambridge.

Corry, Richard (2006). 'Causal Realism and the Laws of Nature'. *Philosophy of Science*, 73(3): 261–76.

Corry, Richard (2011). 'Can Dispositional Essences Ground the Laws of Nature?' *Australasian Journal of Philosophy*, 89(2): 263–75. doi:10.1080/ 00048401003660325.

Corry, Richard (2013). 'Emerging from the Causal Drain'. *Philosophical Studies*, 165(1): 29–47. doi:10.1007/s11098-012-9918-3.

Crane, Tim (2001). 'The Significance of Emergence'. In Barry Loewer and Carl Gillett (editors), *Physicalism and its Discontents*, pp. 207–24. Cambridge University Press, Cambridge. https://philpapers.org/archive/CRATSO-13.pdf.

Creary, Lewis (1981). 'Causal Explanation and the Reality of Natural Component Forces'. *Pacific Philosophical Quarterly*, 62(2): 148–57.

Crisp, Thomas M. and Warfield, Ted A. (2001). 'Kim's Master Argument. A Review of Jaegwon Kim, Mind in a Physical World'. *Noûs*, 35(2): 304–16. doi:10.1111/0029-4624. 00299.

Dalibard, J., Dupont-Roc, J., and Cohen-Tannoudji, C. (1982). 'Vacuum Fluctuations and Radiation Reaction: Identification of their Respective Contributions'. *Journal de Physique*, 43(11): 1617–38. doi:10.1051/jphys: 0198200430110161700.

Darden, Lindley (2002). 'Strategies for Discovering Mechanisms: Schema Instantiation, Modular Subassembly, Forward/Backward Chaining'. *Philosophy of Science*, 69(S3): S354–S365. doi:10.1086/341858.

Davies, Paul (2006). 'The Physics of Downward Causation'. In Philip Clayton and Paul Davies (editors), *The Re-Emergence of Emergence*, pp. 35–52. Oxford University Press.

Dawkins, Richard (1986). *The Blind Watchmaker: Why the Evidence of Evolution Reveals a Universe without Design*. Norton, New York.

Descartes, Rene (1911). *The Philosophical Works of Descartes*. Cambridge University Press, London. https://archive.org/details/philosophicalwor01desc.

Dowe, Phil (1992). 'Wesley Salmon's Process Theory of Causality and the Conserved Quantity Theory'. *Philosophy of Science*, 59: 195–216.

Dowe, Phil (2000). *Physical Causation*. Cambridge Studies in Probability, Induction, and Decision Theory. Cambridge University Press, Cambridge and New York.

Dretske, Fred (1977). 'Laws of Nature'. *Philosophy of Science*, 44(2): 248–68.

Ducasse, C. J. (1926). 'On the Nature and the Observability of the Causal Relation'. *The Journal of Philosophy*, 23(3): 57–68. doi:10.2307/2014377.

Dumsday, Travis (2013). 'Laws of Nature Don't Have Ceteris Paribus Clauses, They are Ceteris Paribus Clauses'. *Ratio*, 26(2): 134–47. doi:10.1111/rati. 12000.

Earman, John (1984). 'Laws of Nature: The Empiricist Challenge'. In Radu J. Bogdan (editor), *D. M. Armstrong*, volume 4 of *Profiles (An International Series on Contemporary Philosophers and Logicians)*, pp. 191–223. Springer, Dordrecht. https://doi.org/10.1007/ 978-94-009-6280-4_8.

Earman, John and Roberts, John (1999). 'Ceteris Paribus, There is No Problem of Provisos'. *Synthese*, 118(3): 439–78.

Eberle, Rolf A. (1967). 'Some Complete Calculi of Individuals'. *Notre Dame Journal of Formal Logic*, 8(4): 267–78. doi:10.1305/ndjfl/1094068838.

Eells, Ellery (1991). *Probabilistic Causality*. Cambridge University Press, Cambridge.

Elliott, William H. and Elliott, Daphne C. (2005). *Biochemistry and Molecular Biology*. Oxford University Press, Oxford and New York, third edition.

Ellis, Brian (2001). *Scientific Essentialism*. Cambridge University Press, Cambridge.

Ellis, Brian and Lierse, Caroline (1994). 'Dispositional Essentialism'. *Australasian Journal of Philosophy*, 72(1): 27–45. doi:10.1080/00048409412345861.

Elser, James J. and Hamilton, Andrew (2007). 'Stoichiometry and the New Biology: The Future is Now'. *PLOS Biology*, 5(7): e181. doi:10.1371/journal. pbio.0050181.

Elton, Charles and Nicholson, Mary (1942). 'The Ten-Year Cycle in Numbers of the Lynx in Canada'. *Journal of Animal Ecology*, 11(2): 215–44. doi:10.2307/ 1358.

Fales, Evan (1990). *Causation and Universals*. Routledge.

Fang, Ferric C. and Casadevall, Arturo (2011). 'Reductionistic and Holistic Science'. *Infection and Immunity*, 79(4): 1401–4. doi:10.1128/IAI.01343-10.

Fara, Michael (2005). 'Dispositions and Habituals'. *Noûs*, 39(1): 43–82.

Feyerabend, Paul K. (1962). 'Explanation, Reduction and Empiricism'. In H. Feigl and G. Maxwell (editors), *Scientific Explanation, Space, and Time*, number III in Minnesota Studies in the Philosophy of Science, pp. 28–97. University of Minnesota Press, Minneapolis.

Field, Hartry (2003). 'Causation in a Physical World'. In M Loux and D Zimmerman (editors), *The Oxford Handbook of Metaphysics*, pp. 435–60. Oxford University Press, Oxford.

Fodor, J. A. (1974). 'Special Sciences (or: The Disunity of Science as a Working Hypothesis)'. *Synthese*, 28(2): 97–115. https://www.jstor.org/stable/20114958.

Forster, Malcolm (1988a). 'The Confirmation of Common Component Causes'. In A Fine and J. Leplin (editors), *PSA 1988*, volume 1, pp. 3–9.

Forster, Malcolm (1988b). 'Unification, Explanation, and the Composition of Causes in Newtonian Mechanics'. *Studies in History and Philosophy of Science*, 19(1): 55–101.

Friedman, Michael (1974). 'Explanation and Scientific Understanding'. *Journal of Philosophy*, 71(1): 5–19.

Gillett, Carl (2016). *Reduction and Emergence in Science and Philosophy*. Cambridge University Press, New York.

Glennan, Stuart (2002). 'Rethinking Mechanistic Explanation'. *Philosophy of Science*, 69(S3): S342–S353. doi:10.1086/341857.

Glennan, Stuart (2017). *The New Mechanical Philosophy*. Oxford University Press, Oxford, New York.

Glennan, Stuart S. (1996). 'Mechanisms and the Nature of Causation'. *Erkenntnis*, 44(1): 49–71. doi:10.1007/BF00172853.

Godfrey-Smith, Peter (2009). 'Causal Pluralism'. In Helen Beebee, Peter Menzies, and Christopher Hitchcock (editors), *The Oxford Handbook of Causation*, pp. 326–37. Oxford University Press, Oxford.

Grassmann, Hermann Günther (1995). *A New Branch of Mathematics: The 'Ausdehnungslehre' of 1844 and Other Works*. Open Court Publishing Company, Chicago.

Gundersen, Lars (2002). 'In Defence of the Conditional Account of Dispositions'. *Synthese*, 130(3): 389–411. http://www.jstor.org/stable/20117224.

Hall, Ned (2004). 'Two Concepts of Causation'. In John Collins, Ned Hall, and Laurie Paul (editors), *Causation and Counterfactuals*, pp. 225–76. MIT Press, Cambridge, MA.

Harré, Rom (1970). 'Powers'. *The British Journal for the Philosophy of Science*, 21(1): 81–101. doi:10.1093/bjps/21.1.81.

Harré, Rom and Madden, Edward H. (1973). 'Natural Powers and Powerful Natures'. *Philosophy*, 48(185): 209–30. http://www.jstor.org/stable/3749407.

Harré, Rom and Madden, Edward H. (1975). *Causal Powers: A Theory of Natural Necessity*. Basil Blackwell, Oxford.

Harsanyi, John C. (1977). 'Rule Utilitarianism and Decision Theory'. *Erkenntnis*, 11(1): 25–53.

Hart, Herbert Lionel Adolphus and Honoré, Tony (1959). *Causation in the Law*. Clarendon Press, Oxford.

Hausman, D. M. and Woodward, J. (1999). 'Independence, Invariance and the Causal Markov Condition'. *The British Journal for the Philosophy of Science*, 50(4): 521–83. doi:10.1093/bjps/50.4.521.

Healey, Richard (1994). 'Nonseparable Processes and Causal Explanation'. *Studies in History and Philosophy of Science Part A*, 25(3): 337–74. doi:10. 1016/0039-3681(94)90057-4.

Hempel, Carl and Oppenheim, Paul (1948). 'Studies in the Logic of Explanation'. *Philosophy of Science*, 15(2): 135–75.

Hitchcock, Christopher (2001). 'The Intransitivity of Causation Revealed in Equations and Graphs'. *The Journal of Philosophy*, 98(6): 273–99. doi:10. 2307/2678432.

Hitchcock, Christopher (2003). 'Of Humean Bondage'. *The British Journal for the Philosophy of Science*, 54(1): 1–25. doi:10.1093/bjps/54.1.1.

Hitchcock, Christopher (2006). 'What's Wrong with Neuron Diagrams?' In J. K. Campbell, M. O'Rourke, and H. Silverstein (editors), *Causation and Explanation*. MIT Press, Cambridge, MA.

Hitchcock, Christopher (2007). 'What Russell Got Right'. In Huw Price and Richard Corry (editors), *Causation, Physics, and the Constitution of Reality: Russell's Republic Revisited*, pp. 45–65. Oxford University Press, Oxford.

Hobson, Art (2011). 'Teaching Elementary Particle Physics, Part I'. *The Physics Teacher*, 49: 12–15. doi:10.1119/1.3527746.

Holland, John Henry (2000). *Emergence: From Chaos to Order*. Oxford University Press, Oxford.

Hooker, Brad (2000). *Ideal Code, Real World: A Rule-Consequentialist Theory of Morality*. Oxford University Press, Oxford.

Hume, David ([1748] 1990). *Enquiries Concerning Human Understanding and Concerning the Principles of Morals*. Clarendon Press, Oxford.

Humphreys, Paul (1997). 'How Properties Emerge'. *Philosophy of Science*, 64(1): 1–17. http://www.jstor.org/stable/188367.

Hüttemann, Andreas (2004). *What's Wrong with Microphysicalism?* International Library of Philosophy. Routledge, London and New York.

Hüttemann, Andreas and Love, Alan C. (2011). 'Aspects of Reductive Explanation in Biological Science: Intrinsicality, Fundamentality, and Temporality'. *British Journal for the Philosophy of Science*, 62(3): 519–49. doi:10.1093/bjps/axr006.

Kaiser, Marie I. (2015). *Reductive Explanation in the Biological Sciences*. Springer, Cham, Heidelberg, New York, Dordrecht, and London.

Kant, Immanuel ([1787] 1929). *Critique of Pure Reason*. Macmillan, Toronto. Translated by Norman Kemp Smith.

Kauffman, Stuart A. (1970). 'Articulation of Parts Explanation in Biology and the Rational Search for Them'. *PSA: Proceedings of the Biennial Meeting of the Philosophy of Science Association*, 1970: 257–72. doi:10.1086/ psaprocbienmeetp.1970.495768.

Kim, Jaegwon (1989). 'Mechanism, Purpose, and Explanatory Exclusion'. *Philosophical Perspectives*, 3: 77–108. http://www.jstor.org/stable/2214264.

Kim, Jaegwon (1990). 'Explanatory Exclusion and the Problem of Mental Causation'. In Enrique Villanueva (editor), *Information, Semantics, and Epistemology*, pp. 36–56. Basil Blackwell, Cambridge, MA.

Kim, Jaegwon (1992). 'Downward Causation' in Emergentism and Non-Reductive Phys-icalism'. In Ansgar Beckermann, Hans Flohr, and Jaegwon Kim (editors), *Emergence or Reduction? Essays on the Prospects of Nonreductive Physicalism*. Walter de Gruyter, Berlin.

Kim, Jaegwon (1993a). 'The Non-Reductivist's Troubles with Mental Causation'. In John Heil and Alfred Mele (editors), *Mental Causation*, pp. 189–210. Clarendon Press, Oxford.

Kim, Jaegwon (1993b). *Supervenience and Mind*. Cambridge University Press, Cambridge.

Kim, Jaegwon (1998). *Mind in a Physical World: An Essay on the Mind-Body Problem and Mental Causation*. Bradford Books, Cambridge, MA.

Kim, Jaegwon (1999). 'Making Sense of Emergence'. *Philosophical Studies*, 95(1–2): 3–36. doi:10.1023/A:1004563122154.

Kim, Jaegwon (2003). 'Blocking Causal Drain and Other Maintenance Chores with Mental Causation'. *Philosophy and Phenomenological Research*, 67(1): 151–76. http://www.jstor.org/stable/20140586.

Kim, Jaegwon (2005). *Physicalism, or Something Near Enough*. Princeton University Press, Princeton, NJ.

Kim, Jaegwon (2006a). 'Being Realistic about Emergence'. In Philip Clayton and Paul Davies (editors), *The Re-Emergence of Emergence*. Oxford University Press, Oxford.

Kim, Jaegwon (2006b). 'Emergence: Core Ideas and Issues'. *Synthese*, 151(3): 547–59. http://dx.doi.org/10.1007/s11229-006-9025-0.

Kim, Jaegwon (2006c). *Philosophy of Mind*. Westview Press, Cambridge, MA, second edition.

Kistler, Max (2002). 'The Causal Criterion of Reality and the Necessity of Laws of Nature'. *Metaphysica*, 3(1): 57–86.

Kistler, Max (2006). 'New Perspectives on Reduction and Emergence in Physics, Biology and Psychology'. *Synthese*, 151(3):311–12. http://dx.doi.org/10.1007/s11229-006-9014-3.

Kitcher, Philip (1989). 'Explanatory Unification and the Causal Structure of the World'. In Philip Kitcher and Wesley Salmon (editors), *Scientific Explanation*, pp. 410–505. University of Minnesota Press, Minneapolis, MN.

Koziolek, Nicholas (2018). 'Belief as the Power to Judge'. *Topoi*. doi:10.1007/ s11245-018-9614-9.

Krebs, Charles J. (2009). *Ecology: The Experimental Analysis of Distribution and Abundance*. Pearson Benjamin Cummings, San Francisco, CA, sixth edition.

Krebs, Charles J., Boonstra, Rudy, Boutin, Stan, and Sinclair, A. R. E. (2001). 'What Drives the 10-Year Cycle of Snowshoe Hares? The Ten-Year Cycle of Snowshoe Hares—One of the Most Striking Features of the Boreal Forest—is a Product of the Interaction between Predation and Food Supplies, as Large-Scale Experiments in the Yukon Have Demon-strated'. *BioScience*, 51(1): 25–35. doi:10.1641/0006-3568(2001)051[0025:WDTYCO] 2.0.CO;2.

Kutach, Douglas (2014). *Causation*. Polity Press, Cambridge.

Latham, Noa (1987). 'Singular Causal Statements and Strict Deterministic Laws'. *Pacific Philosophical Quarterly*, 68: 29–43.

Le Guen, Olivier, Samland, Jana, Friedrich, Thomas, Hanus, Daniel, and Brown, Penelope (2015). 'Making Sense of (Exceptional) Causal Relations. A Cross-Cultural and Cross-Linguistic Study'. *Cognitive Science*, p. 1645. doi:10.3389/fpsyg.2015.01645.

Levins, Richard and Lewontin, Richard (1980). 'Dialectics and Reductionism in Ecology'. *Synthese*, 43(1): 47–78. doi:10.1007/BF00413856.

Lewes, George Henry (1874). *Problems of Life and Mind*, volume II. Trübner & Co, London.

Lewis, David (1973a). 'Causation'. *Journal of Philosophy*, 70(17): 556–67.

Lewis, David (1973b). *Counterfactuals*. Blackwell, Oxford.

Lewis, David (1983a). 'New Work for a Theory of Universals'. *Australasian Journal of Philosophy*, 61(4): 343–77.

Lewis, David (1983b). *Philosophical Papers : Volume I*. Oxford University Press, Oxford.

Lewis, David (1994a). 'Humean Supervenience Debugged'. *Mind*, 103(412): 473–90.

Lewis, David (1994b). 'Reduction of Mind'. In Samuel Guttenplan (editor), *A Companion to the Philosophy of Mind*. Wiley-Blackwell, Oxford.

Lewis, David (1997). 'Finkish Dispositions'. *Philosophical Quarterly*, 47(187): 143–58.

Lewis, David (2000). 'Causation as Influence'. *The Journal of Philosophy*, 97(4): 182–97. doi:10.2307/2678389.

Lewis, Peter (2017). 'Quantum Mechanics, Emergence, and Fundamentality'. *Philosophica*, 91(2): 53–75. http://www.philosophica.ugent.be/fulltexts/92-3.pdf.

Loewer, Barry (1996). 'Humean Supervenience'. *Philosophical Topics*, 24(1): 101–27.

Longworth, Francis (2010). 'Cartwright's Causal Pluralism: A Critique and an Alternative'. *Analysis*, 70(2): 310–18. doi:10.1093/analys/anp109.

Love, Alan C. and Hüttemann, Andreas (2011). 'Comparing Part–Whole Reductive Explanations in Biology and Physics'. In Dennis Dieks, Wenceslao Gonzalo, Thomas Uebel, Stephan Hartmann, and Marcel Weber (editors), *Explanation, Prediction, and Confirmation*, pp. 183–202. Springer, Dordrecht.

Lowe, E. J. (2011). 'How Not to Think of Powers: A Deconstruction of the 'Dispositions and Conditionals' Debate'. *The Monist*, 94(1): 19–33. https://www.jstor.org/stable/23039034.

Lyons, David (1965). *Forms and Limits of Utilitarianism*. Clarendon Press, Oxford.

Machamer, Peter, Darden, Lindley, and Craver, Carl F. (2000). 'Thinking about Mechanisms'. *Philosophy of Science*, 67(1): 1–25. doi:10.1086/392759.

Machery, Edouard (2011). 'Thought Experiments and Philosophical Knowledge'. *Metaphilosophy*, 42(3): 191–214. doi:10.1111/j.1467-9973.2011.01700.x.

Mackie, J. L. (1974). *The Cement of the Universe: A Study of Causation*. Clarendon Press, Oxford.

Maksimovich-Binno, Tiyana (2015). 'Wake up World'. https://www.facebook.com/joinwakeupworld/posts/10206151249080906.

Malzkorn, Wolfgang (2000). 'Realism, Functionalism and the Conditional Analysis of Dispositions'. *Philosophical Quarterly*, 50(201): 452–69.

Manley, David and Wasserman, Ryan (2008). 'On Linking Dispositions and Conditionals'. *Mind*, 117(465): 59–84. doi:10.1093/mind/fzn003.

Marmodoro, Anna (2016). 'Dispositional Modality vis-à-vis Conditional Necessity'. *Philosophical Investigations*, 39(3): 205–14. doi:10.1111/phin.12125.

Marmodoro, Anna (2017). 'Aristotelian Powers at Work: Reciprocity without Symmetry in Causation'. In Jonathan D. Jacobs (editor), *Causal Powers*, pp. 57–76. Oxford University Press, Oxford and New York.

Martin, C. B. (2008). *The Mind in Nature*. Oxford University Press, Oxford and New York.

Massin, Olivier (2017). 'The Composition of Forces'. *British Journal for the Philosophy of Science*, 68(3): 805–46. doi:10.1093/bjps/axv048.

Matthiopoulos, Jason (2011). *How to Be a Quantitative Ecologist: The 'A to R' of Green Mathematics and Statistics*. Wiley, Chichester.

Maudlin, Tim (2007). *The Metaphysics within Physics*. Oxford University Press, Oxford.

McDermott, Michael (2002). 'Causation: Influence versus Sufficiency'. *The Journal of Philosophy*, 99(2): 84–101. doi:10.2307/3655553.

McIntyre, Alison (2014). 'Doctrine of Double Effect'. In Edward N. Zalta (editor), *The Stanford Encyclopedia of Philosophy*. Metaphysics Research Lab, Stanford University, winter 2014 edition. https://plato.stanford.edu/archives/win2014/entries/double-effect/.

McKitrick, Jennifer (2010). 'Manifestations as Effects'. In Anna Marmodoro (editor), *The Metaphysics of Powers: Their Grounding and their Manifestations*, pp. 73–83. Routledge, New York.

McLaughlin, Brian P (1992). 'The Rise and Fall of British Emergentism'. In Ansgar Beckermann, Hans Flohr, and Jaegwon Kim (editors), *Emergence or Reduction? Essays on the Prospects of Nonreductive Physicalism*, p. 49. De Gruyter, Berlin.

McLeish, Tom C. B. (2017). 'Strong Emergence and Downward Causation in Biological Physics'. *Philosophica*, 91(2): 113–38. http://www.philosophica.ugent.be/fulltexts/92-5.pdf.

Mellor, D. H. (1974). 'In Defense of Dispositions'. *Philosophical Review*, 83(2): 157–81.

Mellor, D. H. (1982). 'Counting Corners Correctly'. *Analysis*, 42(2): 96–7.

Mellor, D. H. (2000). 'The Semantics and Ontology of Dispositions'. *Mind*, 109(436): 757–80.

Mendel, Gregor (1866). 'Versuche über Pflanzen-Hybriden'. In *Verhandlungen des naturforschenden Vereines in Brünn, Band IV für das Jahr 1865*, pp. 3–47. Verlage des Vereines, Brünn. https://www.biodiversitylibrary.org/item/124139.

Menzies, Peter (2007). 'Causation in Context'. In Huw Price and Richard Corry (editors), *Causation, Physics, and the Constitution of Reality: Russell's Republic Revisited*, pp. 191–223. Oxford University Press, Oxford.

Menzies, Peter and Price, Huw (1993). 'Causation as a Secondary Quality'. *British Journal for the Philosophy of Science*, 44(2): 187–203. http://www.jstor.org/stable/687643.

Merricks, Trenton (2001). *Objects and Persons*. Oxford University Press, Oxford.

Mill, John Stewart (1843). *A System of Logic*. Longmans, Green Reader, and Dyer, London, seventh edition.

Molnar, George (2003). *Powers*. Oxford University Press, Oxford.

Morgan, C. Lloyd (1923). *Emergent Evolution*. Williams and Norgate, London.

Mumford, Stephen (1998). *Dispositions*. Oxford University Press, Oxford.

Mumford, Stephen (2004). *Laws in Nature*. Routledge, Abingdon.

Mumford, Stephen and Anjum, Rani Lill (2010). 'A Powerful Theory of Causation'. In Anna Marmodoro (editor), *The Metaphysics of Powers: Their Grounding and their Manifestations*, pp. 143–159. Routledge, New York.

Mumford, Stephen and Anjum, Rani Lill (2011a). *Getting Causes from Powers*. Oxford University Press, Oxford.

Mumford, Stephen and Anjum, Rani Lill (2011b). 'Spoils to the Vector: How to Model Causes if You Are a Realist about Powers'. *The Monist*, 94(1): 54–80. http://www.jstor.org/stable/23039036.

Nagel, Ernst (1949). 'The Meaning of Reduction in the Natural Sciences'. In R. C. Stauffer (editor), *Science and Civilization*, pp. 99–135. University of Wisconsin Press, Madison, WI.

Nagel, Ernst (1961). *The Structure of Science. Problems in the Logic of Explanation*. Harcourt, New York.

Norton, John D (2007). 'Causation as Folk Science'. In Huw Price and Richard Corry (editors), *Causation, Physics, and the Constitution of Reality: Russell's Republic Revisited*, pp. 11–44. Oxford University Press, Oxford.

O'Connor, Timothy (1994). 'Emergent Properties'. *American Philosophical Quarterly*, 31(2): 91–104. http://www.jstor.org/stable/20014490.

O'Connor, Timothy (2000). *Persons and Causes: The Metaphysics of Free Will*. Oxford University Press.

O'Connor, Timothy (2009). 'Agent-Causal Power'. In Toby Handfield (editor), *Dispositions and Causes*. Oxford University Press, Oxford.

O'Connor, Timothy and Wong, Hong Yu (2005). 'The Metaphysics of Emergence'. *Noûs*, 39(4): 658–78. doi:10.1111/j.0029-4624.2005.00543.x.

O'Connor, Timothy and Wong, Hong Yu (2015). 'Emergent Properties'. In Edward N. Zalta (editor), *The Stanford Encyclopedia of Philosophy*. Summer 2015 edition. http://plato.stanford.edu/archives/sum2015/entries/properties-emergent/.

OpenStax (2016). *Anatomy and Physiology*. OpenStax CNX. http://cnx.org/contents/14fb4ad7-39a1-4eee-ab6e-3ef2482e3e22@8.24.

Pearl, Judea (2000). *Causality*. Cambridge University Press, New York.

Peskin, Michael E. and Schroeder, Daniel V. (1995). *An Introduction to Quantum Field Theory*. Perseus Books, Reading, MA.

Polanyi, Michael (1968). 'Life's Irreducible Structure: Live Mechanisms and Information in DNA Are Boundary Conditions with a Sequence of Boundaries above Them'. *Science*, 160(3834): 1308–12. doi:10.1126/science.160.3834.1308.

Price, Donald D., McGrath, Patricia A., Rafii, Amir, and Buckingham, Barbara (1983). 'The Validation of Visual Analogue Scales as Ratio Scale Measures for Chronic and Experimental Pain'. *Pain*, 17(1): 45–56. doi:10. 1016/0304-3959(83)90126-4.

Prior, Arthur (1985). *Dispositions*. Aberdeen University Press, Aberdeen.

Psillos, Stathis (2010). 'Causal Pluralism'. In Robrecht Vanderbeeken and Bart D'Hooghe (editors), *Worldviews, Science, and Us: Studies of Analytical Metaphysics: A Selection of Topics from a Methodological Perspective*, pp. 131–51. World Scientific, Hackensack, NJ.

Putnam, Hilary (1967). 'Psychological Predicates'. In W. H. Capitan and D. D. Merrill (editors), *Art, Mind, and Religion*, pp. 37–48. University of Pittsburgh Press, Pittsburgh, PA.

Ramsey, Frank (1978). *Foundations*. Routledge and Kegan Paul, London.

Reutlinger, Alexander (2012). 'Getting Rid of Interventions'. *Studies in History and Philosophy of Science Part C*, 43(4): 787–95.

Russell, Bertrand (1903). *The Principles of Mathematics*. Routledge, London.

Russell, Bertrand (1913). 'On the Notion of Cause'. *Proceedings of the Aristotelian Society*, 13: 1–26. http://www.jstor.org/stable/4543833.

Ryle, Gilbert (1949). *The Concept of Mind*. Hutchinson and Co, London.

Salmon, Wesley (1984). *Scientific Explanation and the Causal Structure of the World*. Princeton University Press, Princeton, NJ.

Salmon, Wesley C. (1997). 'Causality and Explanation: A Reply to Two Critiques'. *Philosophy of Science*, 64(3): 461–77. http://www.jstor.org/stable/188320.

Sarkar, Sahotra (1998). *Genetics and Reductionism*. Cambridge University Press, Cambridge.

Schaffer, Jonathan (2005). 'Contrastive Causation'. *Philosophical Review*, 114(3): 327–58.

Schaffer, Jonathan (2014). 'The Metaphysics of Causation'. *The Stanford Encyclopedia of Philosophy* (Summer 2014 Edition), Edward N. Zalta (ed), http:// plato.stanford.edu/archives/sum2014/entries/causation-metaphysics/.

Schaffner, Kenneth (2006). 'Reduction: The Cheshire Cat Problem and a Return to Roots'. *Synthese*, 151(3): 377–402. http://dx.doi.org/10.1007/s11229-006-9031-2.

Schaffner, Kenneth F. (1993). *Discovery and Explanation in Biology and Medicine*. Science and its Conceptual Foundations. University of Chicago Press, Chicago.

Schumacher, Ernst Friedrich (1979). *Good Work*. Jonathan Cape, London.

Shoemaker, Sydney (1980). 'Causality and Properties'. In Peter van Inwagen (editor), *Time and Cause: Essays Presented to Richard Taylor*, pp. 109–36. D. Reidel Publishing, Dordrecht.

Shoemaker, Sydney (2002). 'Kim on Emergence'. *Philosophical Studies*, 108(1–2): 53–63. doi:10.1023/A:1015708030227.

Sider, Ted (2001). *Four-Dimensionalism*. Clarendon Press, Oxford.

Skyrms, Brian (1984). 'EPR: Lessons for Metaphysics'. *Midwest Studies In Philosophy*, 9(1): 245–55. doi:10.1111/j.1475-4975.1984.tb00062.x.

Smith, Sheldon (2002). 'Violated Laws, Ceteris Paribus Clauses, and Capacities'. *Synthese*, 130(2): 235–64. http://www.jstor.org/stable/20117215.

Sober, Elliott (1984). 'Two Concepts of Cause'. *PSA: Proceedings of the Biennial Meeting of the Philosophy of Science Association* 1984: 405–24.

Sober, Elliott (1988). 'Apportioning Causal Responsibility'. *The Journal of Philosophy*, 85(6): 303–18. doi:10.2307/2026721.

Sperry, Roger W. (1969). 'A Modified Concept of Consciousness'. *Psychological Review*, 76(6): 532–6. doi:10.1037/h0028156.

Spinoza, Benedict de (1677). *Ethics*. Penguin Classics, London and New York.

Spirtes, Peter, Glymour, Clark, and Scheines, Richard (1993). *Causation, Prediction, and Search*. Springer, New York. http://dx.doi.org/10.1007/978-1-4612-2748-9.

Stenseth, Nils C., Falck, Wilhelm, Bjørnstad, Ottar N., and Krebs, Charles J. (1997). 'Population Regulation in Snowshoe Hare and Canadian Lynx: Asymmetric Food Web Configurations between Hare and Lynx'. *Proceedings of the National Academy of Sciences*, 94(10): 5147–52. doi:10.1073/pnas.94. 10.5147.

Suppes, Patrick (1970). *A Probabilistic Theory of Causality*. Acta Philosophica Fennica. North-Holland, Amsterdam.

Swoyer, Chris (1982). 'The Nature of Natural Laws'. *Australasian Journal of Philosophy*, 60(3): 203–23.

Tooley, Michael. (1977). 'The Nature of Laws'. *Canadian Journal of Philosophy*, 7(4): 667–98.

Tooley, Michael (1987). *Causation: A Realist Approach*. Oxford University Press.

Townsend, Colin R., Begon, Michael, and Harper, John L. (2008). *Essentials of Ecology*. Blackwell Publishers, Malden, MA, third edition.

Urry, Lisa A, Meyers, Noel, Cain, Michael L., Wasserman, Steven A., Minorsky, Peter V., and Reece, Jane B. (2018). *Campbell Biology: Australia and New Zealand Edition*. Pearson, Melbourne, eleventh edition.

Van Cleve, James (1990). 'Mind-Dust or Magic? Panpsychism versus Emergence'. *Philosophical Perspectives*, 4: 215–26. doi:10.2307/2214193.

Van Fraassen, Bas C. (1989). *Laws and Symmetry*. Oxford University Press, Oxford and New York.

Van Gulick, Robert (2001). 'Reduction, Emergence and Other Recent Options on the Mind/Body Problem: A Philosophic Overview'. *Journal of Consciousness Studies*, 8(9–10): 1–34.

Van Inwagen, Peter (1990). *Material Beings*. Cornell University Press, Ithaca, NY.

Van Regenmortel, Marc H.V. (2004). 'Reductionism and Complexity in Molecular Biology'. *EMBO Reports*, 5(11): 1016–20. doi:10.1038/sj.embor. 7400284.

Williams, Bernard (1973). 'A Critique of Utilitarianism'. In J. J. C. Smart and Bernard Williams (editors), *Utilitarianism: For and against*, pp. 77–150. Cambridge University Press, Cambridge.

Wilson, Alex (2016). 'Victim-Blaming the Left is so Reductionist and Dumb'. https://www.smaqtalk.com/chat/discussion/1-196488-victim-blaming-the-left-is-so-reductionist-and-dumb.html.

Wilson, Jessica (2009). 'The Causal Argument Against Component Forces'. *Dialectica*, 63(4): 525–54. doi:10.1111/j.1746-8361.2009.01216.x.

Wilson, Jessica (2015). 'Metaphysical Emergence: Weak and Strong'. In Tomasz Bigaj and Christian Wuthrich (editors), *Metaphysics in Contemporary Physics*, pp. 251–306. Poznan Studies in the Philosophy of the Sciences and the Humanities.

Wimsatt, William and Sarkar, Sahotra (2006). 'Reductionism'. In Sahotra Sarkar and Jessica Pfeifer (editors), *The Philosophy of Science: An Encyclopedia*, pp. 696–702. Routledge, New York.

Wimsatt, William C. (1976). 'Reductive Explanation: A Functional Account'. In R. S. Cohen, C. A. Hooker, A. C. Michalos, and J. W. Van Evra (editors), *PSA 1974. Boston Studies in the Philosophy of Science*, volume 32, pp. 671–710. Springer, Dordrecht. doi:10.1007/978-94-010-1449-6_38.

Wimsatt, William C. (2006). 'Reductionism and its Heuristics: Making Methodological Reductionism Honest'. *Synthese*, 151(3): 445–75. doi:10.1007/ s11229-006-9017-0.

Wong, Hong Yu (2010). 'The Secret Lives of Emergents'. In Antonella Corradini and Timothy O'Connor (editors), *Emergence in Science and Philosophy*, pp. 7–24. Routledge, New York.

Woodward, James (2003). *Making Things Happen: A Theory of Causal Explanation*. Oxford University Press, Oxford.

Woodward, Jim (2002). 'What is a Mechanism? A Counterfactual Account'. *Philosophy of Science*, 69(S3): S366–S377. doi:10.1086/341859.

Yablo, Stephen (1992). 'Mental Causation'. *The Philosophical Review*, 101(2): 245–80. doi:10.2307/2185535.

Yokoyama, Hiroyuki and Ujihara, Kikuo (editors) (1995). *Spontaneous Emission and Laser Oscillation in Microcavities*. CRC Press, Boca Raton, FL.

Index